Operation and Maintenance of Farm Machinery

CONTENT VIBES
PUBLISHERS • PRINTERS • AGGREGATORS

About the Author

Tejas Ashok Bhosale completed his Bachelor of Technology (B. Tech.) from Mahatma Phule Krishi Vidyapeeth (MPKV), Rahuri, Maharashtra. He qualified for the Junior Research Fellowship (JRF) examination conducted by the Indian Council of Agricultural Research (ICAR). Subsequently, he earned his Master of Technology (M. Tech.) from G.B. Pant University of Agriculture and Technology (GBPUAT), Pantnagar. He has also qualified for the ICAR Agricultural Scientists Recruitment Board National Eligibility Test (ICAR-ASRB-NET). With over 12 years of teaching experience in State Agricultural Universities (SAUs), Tejas is currently serving as an Assistant Professor at Veer Chandra Singh Garhwali Uttarakhand University of Horticulture and Forestry (VCSGUUHF), Bharsar, Pauri Garhwal, Uttarakhand. He is a DGCA-certified drone pilot, specializing in the application of drones within the agricultural sector. Tejas has published 7 research papers, 5 popular articles, 1 book chapter, and 1 book. He has handled one project as Principal Investigator (PI) and another as Co-Principal Investigator (Co-PI). Additionally, he possesses administrative experience in various capacities within SAUs. His areas of specialization include agricultural mechanization, precision agriculture, the use of drones in agriculture, and the design and development of intercultural and harvesting machinery. Tejas's expertise and contributions significantly enhance agricultural technology and education.

Operation and Maintenance of Farm Machinery

Tejas Ashok Bhosale
Assistant Professor
Agricultural Engineering
Department of Natural Resource Management
College of Horticulture
Veer Chandra Singh Garhwali Uttarakhand University of Horticulture and Forestry
Bharsar, Pauri Garhwal, Uttarakhand

CV CONTENT VIBES
PUBLISHERS • PRINTERS • AGGREGATORS

CONTENT VIBES
PUBLISHERS • PRINTERS • AGGREGATORS

1168, Sector 13, Urban Estate
Karnal-132 001, Haryana
Tel: 91-84470 75807, 18440 41168
Email: contentvibesppa@gmail.com
www.contentvibes.in

Print ISBN: 978-93-67553-55-8

ebooks ISBN: 978-93-67551-4-00

Preface

The advent of mechanization has revolutionized modern farming, enhancing productivity, reducing labor, and enabling farmers to cultivate larger areas of land efficiently. Farm machinery plays a crucial role in ensuring the smooth operation of farming practices, from planting and cultivating to harvesting and irrigation. The evolution of agricultural technology has brought about sophisticated equipment such as tractors, harvesters, plows, seeders, and irrigation systems, which are essential in meeting the growing demands for food and other agricultural products.

However, the mere presence of these machines does not guarantee optimal performance. Proper operation and maintenance are pivotal to ensuring that farm machinery delivers its maximum potential. Operation involves the skillful handling of the equipment, understanding its functions, and ensuring safety and efficiency during use. Maintenance, on the other hand, involves a series of actions aimed at keeping the machinery in good working condition, preventing breakdowns, and extending its lifespan. Both operation and maintenance are vital for reducing downtime and minimizing repair costs, which can otherwise be substantial for farmers.

A major part of machinery operation is understanding the specific requirements and capabilities of each piece of equipment. For example, a tractor might be used for tasks as varied as plowing, hauling, or mowing, each of which requires different operational techniques. Understanding how to operate the machinery according to these tasks ensures better results, reduces fuel consumption, and minimizes wear and tear on the equipment.

Maintenance is equally important and encompasses preventive and corrective measures. Preventive maintenance includes routine checks, cleaning, oil changes, and component inspections that can be scheduled to avoid unexpected failures. Corrective maintenance, in contrast, is performed when something goes wrong, requiring immediate repairs to restore equipment functionality. Regular maintenance schedules can significantly reduce the likelihood of mechanical failures, enhance performance, and ensure that machines operate safely.

The ability to properly maintain farm machinery is not just a technical skill; it is also a critical aspect of farm management. Farmers who invest time and effort into understanding the operation and maintenance of their machinery are more likely to achieve higher productivity levels and increase their return on investment. Moreover, as technology continues to advance, staying informed about new developments in machinery and maintenance practices will be essential for farmers to stay competitive and sustainable.

In this context, the importance of proper operation and maintenance cannot be overstated. It is essential not only for the longevity of the machinery but also for the profitability and efficiency of farming operations. The book will explore the best practices for operating and maintaining farm machinery, covering key concepts, tools, and techniques to optimize farm productivity.

—Author

Contents

1

Primary Tillage Equipment Mouldboard Plough and Disc Plough

OVERVIEW OF PRIMARY TILLAGE EQUIPMENT

Primary tillage is the first and one of the most important steps in soil preparation for agriculture. It involves the mechanical disturbance of the soil to break it up, loosen it, and turn it over, creating optimal conditions for crop growth. The aim of primary tillage is to ready the soil for sowing by forming a seedbed, enhancing soil structure, managing weeds, incorporating crop residues, and improving water absorption. Loosening the soil also promotes aeration, which is crucial for root growth and microbial activity. The most common tools that are used for primary tillage are the mouldboard plough and the disc plough, each suited for different soil types and field conditions. These tools are designed to break and loosen the soil, ensuring proper root development and preparation for subsequent stages of cultivation.

The mouldboard plough, one of the oldest and most widely used primary tillage tools, includes several components such as a frame, shares (blades), mouldboards, landsides, and handles. The shares cut into the soil, while the mouldboards flip the soil over, creating furrows and burying crop residues. It is best for deep tillage in lighter, well-drained soils and is generally used in areas with low weed pressure. The depth and width of the cut can be adjusted to suit the soil type and crop requirements.

In contrast, the disc plough uses rotating discs to cut into the soil. Unlike the mouldboard plough, which fully turns the soil, the disc plough slices and fragments it into smaller pieces. This makes it particularly effective in dense, compacted, or rocky soils. Disc ploughs are ideal for regions with heavy, sticky soils or where crop residues are plentiful. The discs cut the soil without fully turning it, which helps reduce soil erosion and preserve moisture levels. Disc ploughs come in single-way and double-way versions, with discs mounted at different angles to improve soil penetration. Both mouldboard and disc ploughs are crucial for primary tillage, but they are suited to different soil conditions. Mouldboard ploughs are more effective in lighter soils that require turning and loosening, while disc ploughs are better for tough, compact soils. Selecting the right plough is essential for preparing the soil adequately, which in turn supports higher crop yields and better soil health. Proper use of these tools also aids in water retention, crop residue management, and weed control, making primary tillage an importabt part of the overall farming process.

Significance of Primary Tillage

Primary tillage is important for several key reasons:

- **Soil Aeration:** It enhances oxygen flow in the soil, which is vital for healthy root development.
- **Weed Control:** Turning the soil uproots and buries weeds, preventing them from regrowing.
- **Incorporating Organic Matter:** Primary tillage incorporates crop residues, manure, and other organic materials into the soil, enriching its quality.
- **Soil Structure Improvement:** It breaks up compacted soil layers, creating space for better water infiltration and root growth.

MOULDBOARD PLOUGH

The mouldboard plough, one of the oldest and most widely used tools for primary tillage, is composed of several essential components: a frame, shares, mouldboards, landsides, and handles. The frame provides structural support for the tool, the shares (or blades) cut into the soil, and the mouldboards turn the soil to form furrows.

Figure 1.1: Mouldboard Plough

The mouldboard plough primarily serves to turn the soil, bury crop residues, and prepare a well-tilled seedbed. It is ideal for deep tillage in lighter soils and is commonly used in areas with minimal weed pressure. The depth and width of the cut can be adjusted to suit specific soil types and crop needs.

Mouldboard Plough are:

- Among the earliest agricultural tools ever developed
- Regarded as the most essential implement for tillage
- Requires more traction energy than any other farming operation
- Cuts, loosens, and inverts the furrow slice, leaving it partially pulverized
- Commonly used to incorporate grass into the soil right after rainfall
- Its design primarily relies on a trial-and-error approach.
- Capable of tackling the most challenging ploughing tasks with excellent penetration efficiency.
- Engineered to perform effectively in all soil types for breaking, raising, and turning soil.
- Suitable for use in stony or root-filled soils.
- Allows quick and easy adjustments, such as adding an extra furrow or repositioning units for additional clearance.
- Features bar points with extended durability, usable until their maximum length.

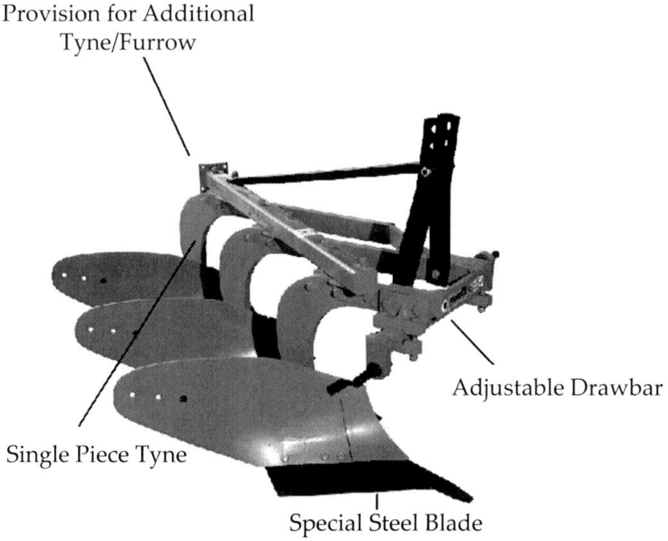

Figure 1.2: Schematic of a Mouldboard Plough

Figure 1.3: Drawbar Assembly

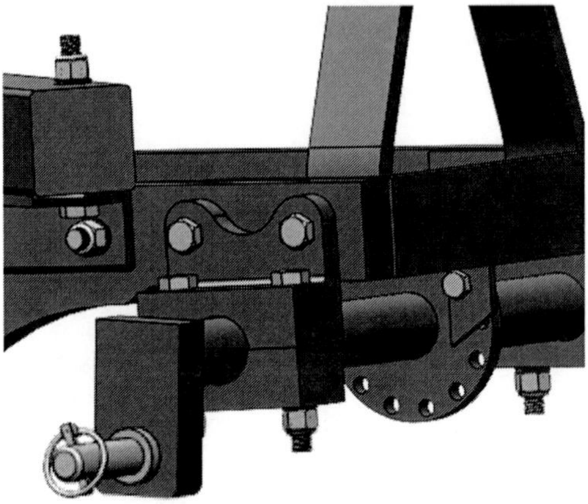

Figure 1.4: Adjustable Drawbar

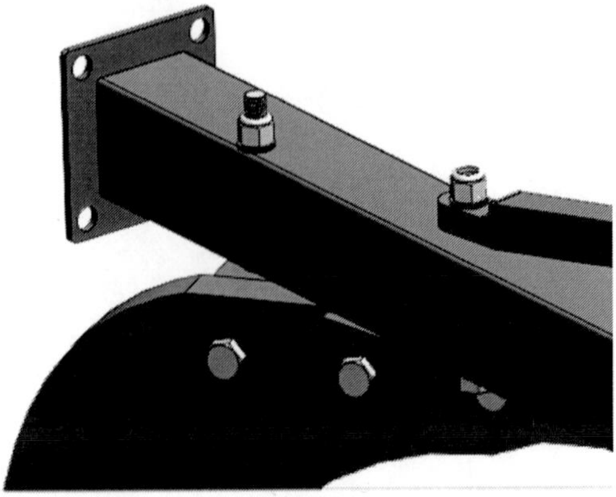

Figure 1.5: Additional Tyne/Furrow

Figure 1.6: 10 mm Thick Blade

Figure 1.7: Single Piece Tyne

Key Components of Mouldboard Plough

Mouldboard

The mouldboard is the primary cutting component of the plough, typically metal and curved to lift and turn the soil.

- **Material:** Made from carbon steel or manganese steel for durability and wear resistance.
- **Shape:** Curved to effectively turn the soil, with designs varying (convex or concave) depending on soil type and tillage depth.
- **Angle of Attack:** The angle at which the mouldboard contacts the soil determines ploughing efficiency and depth.

Share

The share is the blade that cuts into the soil, creating a furrow for the moldboard to lift and turn.

- **Material:** Constructed from high-carbon or alloy steel for hardness and durability.
- **Shape:** Sharp-edged and angled for easy soil penetration and smooth operation.

Beam

The beam serves as the main structural element, connecting the moldboard and share to the hitch or plough frame, ensuring stability.

Handles or Hitch

These points attach the plough to the tractor, enabling it to pull the plough through the soil.

- **Type:** Often a three-point linkage system for secure attachment.
- **Material:** Made from durable steel.

Landside

The landside is a vertical plate that provides stability and guides the plough through the soil.

- **Material:** Typically made of steel.
- **Function:** Ensures the plough remains aligned in the furrow and prevents lateral drifting.

Depth Wheel

Some moldboard ploughs include a depth wheel to adjust the working depth of the plough in the soil.

Depth Control Mechanism

Adjustable depth controls, operated via levers, hydraulics, or mechanical systems, allow the operator to modify the furrow's depth.

- **Hydraulic or Mechanical:** Modern ploughs often feature hydraulic systems for automatic depth adjustments.

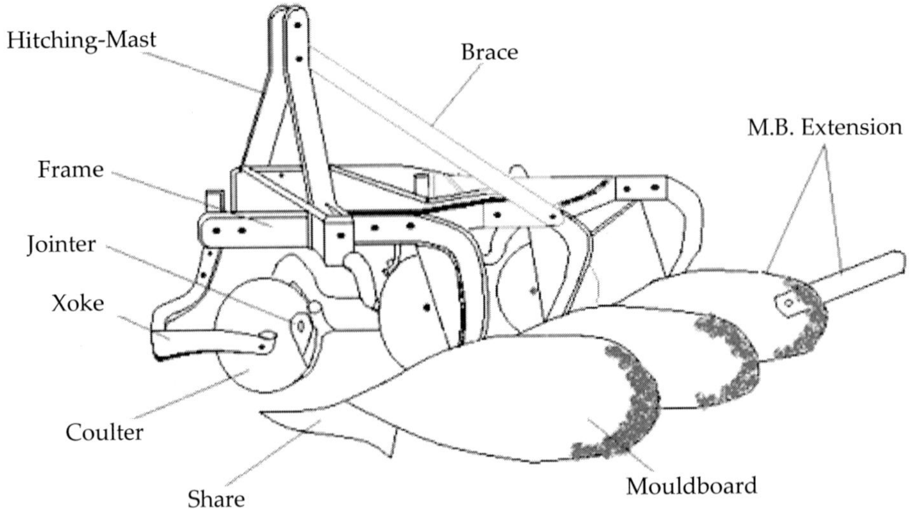

Figure 1.8: A Detailed Schematic of Mouldboard Plough

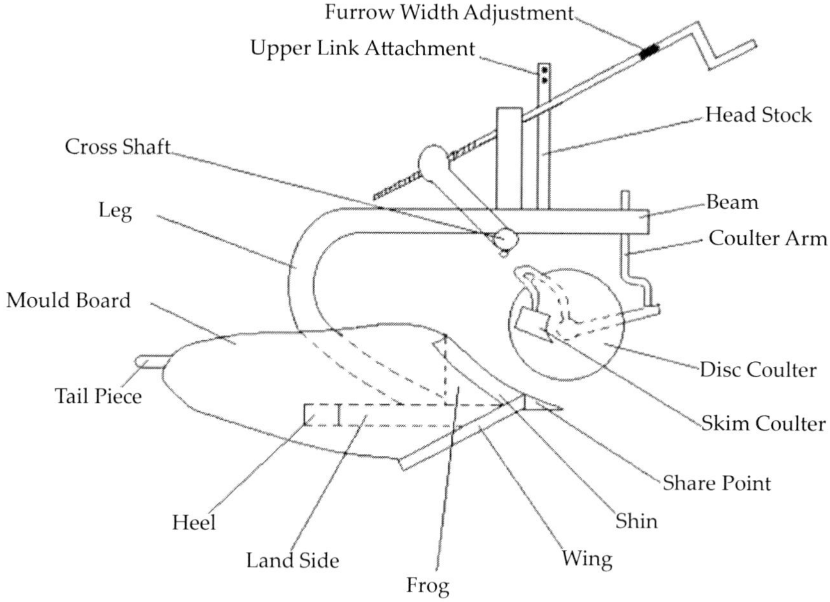

Figure 1.9: Parts of a Mounted Mouldboard Plough

Plow Bottom

The plow bottom is the core component of the plough, shaped like a three-sided wedge. It comprises three key parts: the mouldboard, landside, and share, all firmly attached to the frog. Its primary role is to cut the furrow slice, break apart the soil, and turn the furrow slice to bury crop residues. The size of the plow bottom corresponds to the width of the furrow it is designed to cut

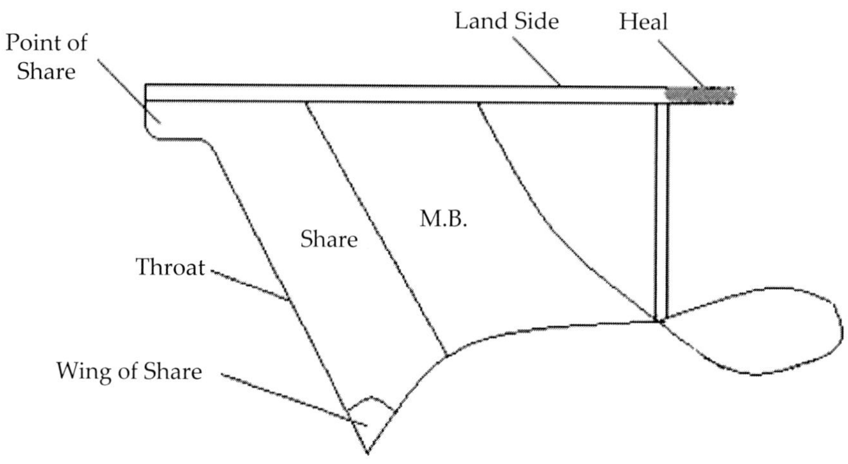

Figure 1.10: Top View of MB Plow Bottom

- **Frog:** It is the base of the plow bottom to which other parts are attached.
- **Tail Piece:** It is the extension of Mouldboard which helps in turning the furrow slice.

Different Variants of Mouldboard

Plow bottoms need to be shaped differently depending on soil conditions. The moisture content and texture of the soil dictate whether the soil should be finely pulverized or simply turned over to be broken down later.

General Purpose and High Speed

This mouldboard has a medium curvature, positioned between the stubble and sod types. It is relatively long with a gradual twist and a slightly convex surface. The slope of the surface is gradual, allowing it to turn a well-defined furrow slice and thoroughly pulverize the soil. These mouldboards are commonly used and suitable for a variety of conditions, typically meeting the general requirements for seedbed preparation.

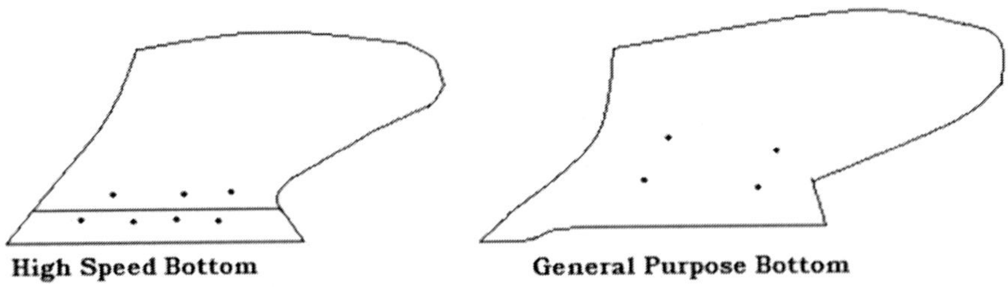

Figure 1.11: General Purpose and High Speed Bottom

Stubble Bottom

This is a shorter, broader mouldboard with a relatively sharp curvature that lifts, breaks, and turns the furrow slice. It is ideal for working in stubble soil that has been under cultivation for several years, where plant residues from previous crops remain on the land during ploughing. This type of mouldboard is not suitable for land heavily covered with grasses. It is typically used on older fields where thorough pulverization is needed. The short, broad design and the abrupt curve near the top provide a higher degree of soil pulverization compared to other types.

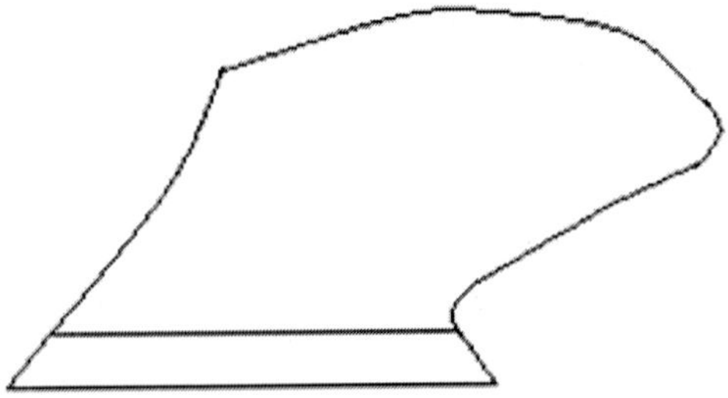

Figure 1.12: Stubble Bottom

Sod or Breaker Bottom

This is a long mouldboard with a gentle curve that lifts and inverts the unbroken furrow slice, effectively turning over densely covered soil. It is particularly useful when the farmer requires complete soil inversion. Designed for use in sod soils (soils with a significant amount of grass), it is ideal for tough soil where furrow slices need to be fully turned to prevent grass from growing. The long, low mouldboard has a gradual spiral twist that inverts the furrow slice with minimal breakup, thoroughly covering the vegetative matter.

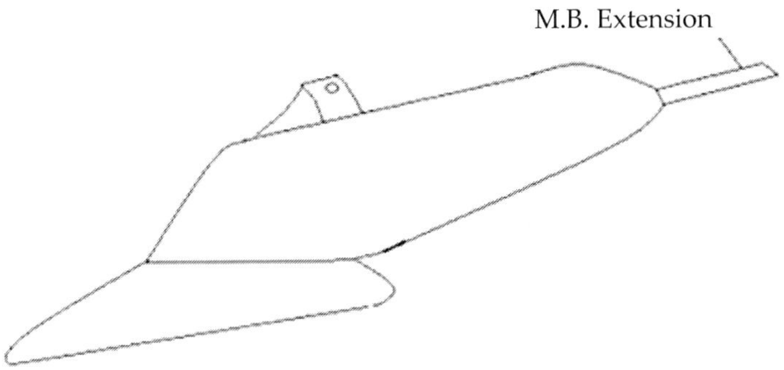

Figure 1.13: Sod or Breaker Bottom

Black Land Bottom

This mouldboard is used for plowing gumbo or buckshot soil, where scouring (soil buildup) is a concern. It features a smaller mouldboard area, and its design helps promote scouring of the soil.

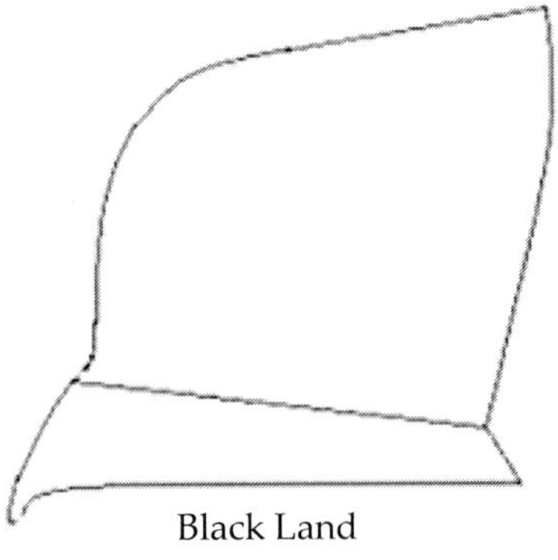

Figure 1.14: Black Land Bottom

Slat Bottom

This mouldboard has a surface made of slats arranged along its length, creating gaps between them. It is commonly used in sticky soils, as a solid mouldboard does not effectively scour in such conditions. Although uncommon, it is particularly beneficial in light and sticky soils where a general-purpose plow may struggle to scour. The slats increase pressure between the soil and the mouldboard, improving the scouring action.

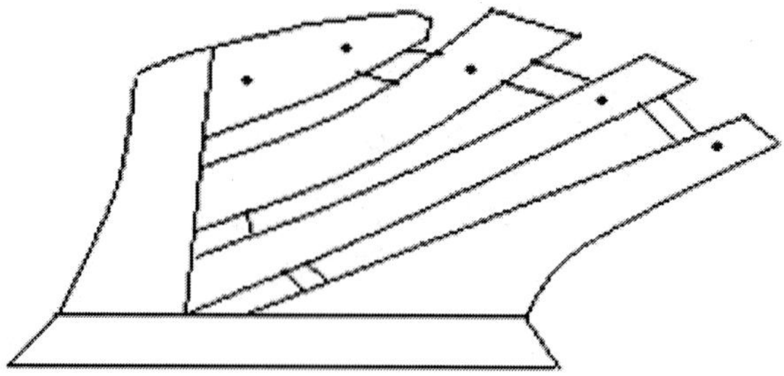

Figure 1.15: Slatted Bottom

Parts of Share

The share is designed to slice through the soil, allowing the plough to work efficiently and break up the compacted ground. Mentioned below are the main parts of the share in a mouldboard plough:

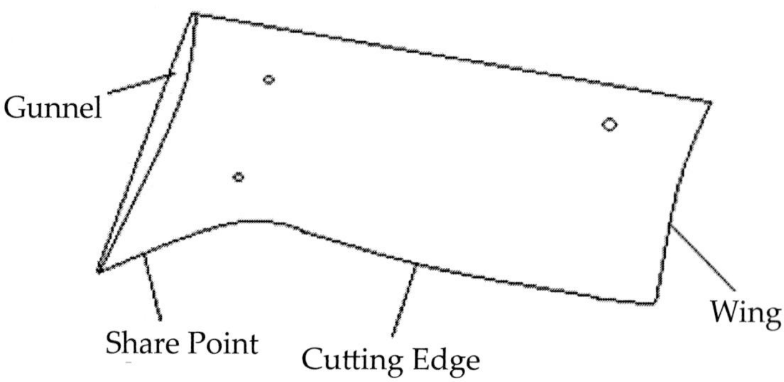

Figure 1.16: Parts of Share.

- **Share Point:** The forward tip of the share that first makes contact with the soil, penetrating it. Typically made of hardened or manganese steel to resist wear from soil friction, it initiates soil penetration and reduces resistance for smoother entry of the plow.
- **Cutting Edge:** The front edge of the share that makes a horizontal cut in the soil. Made from hardened steel for durability and sharpness, its primary role is to slice through the soil, easing the plow's forward movement and aiding soil turnover.
- **Wing of Share:** The outer part of the cutting edge of the share, supporting the plow bottom.
- **Gunnels:** The vertical surface of the share that slides along the furrow wall, absorbing the lateral thrust of the soil and providing support to the plow bottom against the furrow wall.

- **Cleavage Edge:** The edge of the share that forms the joint between the mouldboard and share on the frog.
- **Wing Bearing:** The level portion of the share wing that acts as a support for the outer corners of the plow bottom.
- **Throat:** The main sections of the share, including the share point, wing, and cutting edge, collectively referred to as the throat. The point is the first part to enter the soil, while the wing is the outermost corner of the cutting edge. The curved portion between the point and wing forms the throat of the share.

Different Types of Shares

Slip Share

It is a jointed part with a curved edge, without additional parts. It is a common type of stock, mainly used by farmers. It is simple in design, but has the disadvantage that the entire part must be replaced if it wears out due to continuous use. It is a common type of stock generally used by farmers because it is simple in design. Its major disadvantage is that entire share has to be replaced if it is worn out due to constant use.

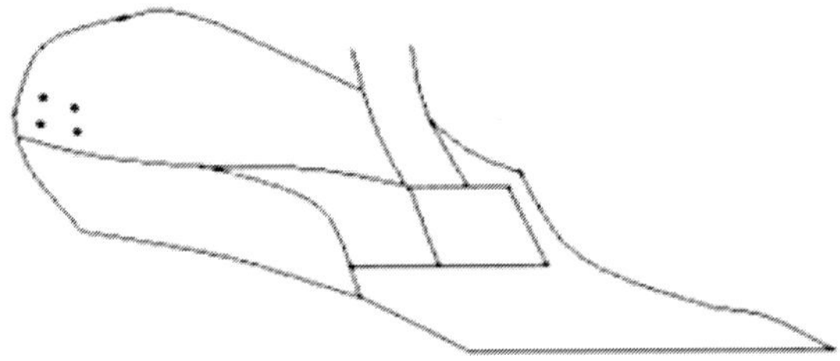

Figure 1.17: Slip Share

Slip-Nose Share

This is a plow whose plow point consists of a small detachable part. The advantage is that the plow point can be replaced as needed. So it is economical

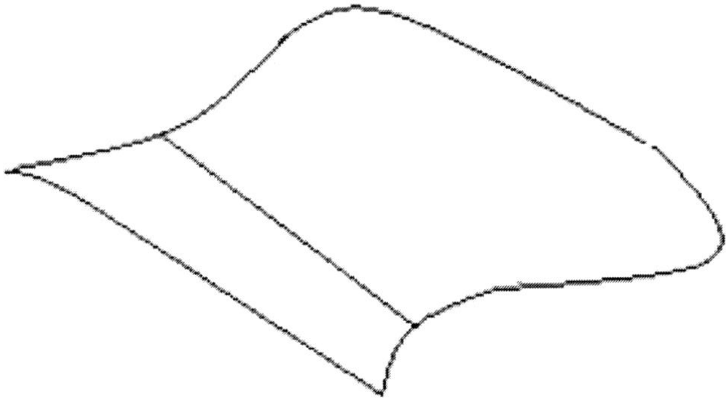

Figure 1.18: Slip Nose Share

Shin Share

It is a share having a shin as an additional part. It is similar to the slip share with the difference that an extension is provided to it by the side of the mouldboard.

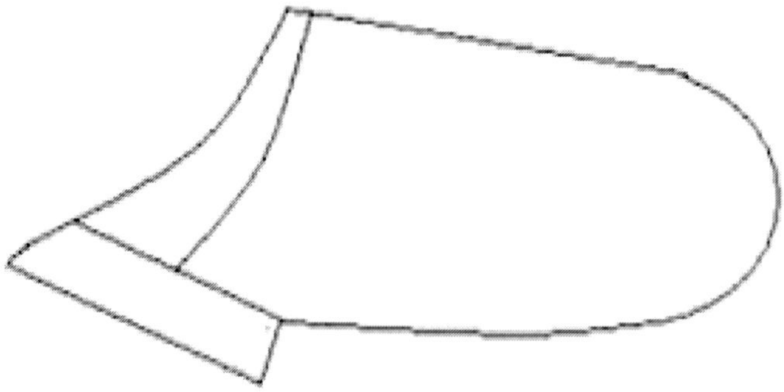

Figure 1.19: Shin Share

Bar Point Share

It is the share in which point of share is provided by an adjustable and replaceable bar. This bar serves the purpose of point of share and landslide of the plow.

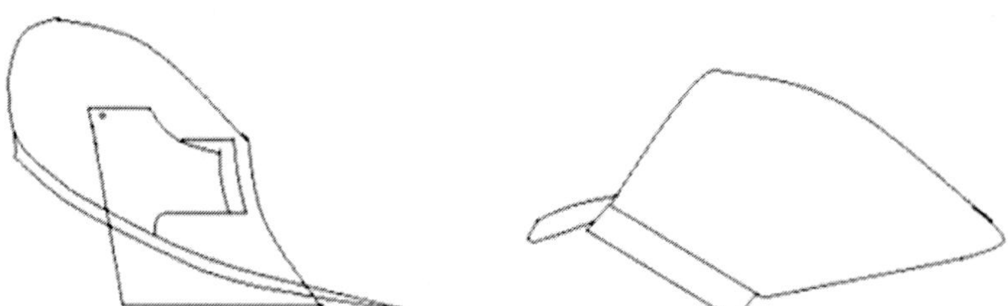

Figure 1.20: Bar Share and Bar Point Share

Materials Utilized for Share

Shares are typically made from chilled cast iron or steel, with the steel containing around 0.70 to 0.80% carbon and 0.50 to 0.80% manganese, along with other minor elements.

- **Cast Iron Shares:** These are used in more affordable plows. They are brittle and can break easily if struck sharply, requiring careful handling. While broken shares need to be replaced, worn ones can be sharpened using a grinding stone.
- **Chilled Cast Iron Shares:** These shares have a longer lifespan due to their resistance to rust and slower wear. They are recommended for use in sandy or hard soils.
- **Soft Center Steel Shares:** These are ideal for soils where soil does not cling to the surface of the share or mouldboard. They have a hard surface and long durability but come at a higher cost.

- **Solid Steel Shares:** Designed for non-abrasive soils, these shares are robust and shock-resistant. They are made from high-carbon steel through a forging process.

Varieties of Mouldboard Plows

- **Trailed Plow:** Also known as a pull-type plow, this is a complete unit supported by two wheels. The entire unit is hitched to the tractor's drawbar and is available in configurations ranging from 1 to 8 bottoms, depending on the tractor's capacity. A single-bottom plow has one bottom, while two to eight bottoms are referred to as gang mouldboard plows. Available sizes include 36, 41, and 46 cm. These plows are less maneuverable.
- **Semi-Mounted Plow:** These plows are more compact and maneuverable compared to pull-type plows. They have the same size and number of bottoms as pull-type plows but are less expensive. Semi-mounted plows exert more vertical load on the tractor's rear wheels, improving tractive ability.
- **Mounted Plow:** Also called direct-mounted, tractor-mounted, or tractor-carried plows, these use the tractor's lift linkages, controlled hydraulically. They are available in 2 to 5 bottoms, depending on the tractor's capacity, with available sizes of 30, 36, and 41 cm. Larger mounted plows can affect tractor stability during transport.
- **One-Way Plow:** This plow turns the soil to the right-hand side. It requires laying out a field in lands, starting with back furrows and ending with dead furrows.
- **Two-Way Plow:** This plow turns the soil to both the right and left. It features two sets of bottoms mounted on a common frame that rotates around a longitudinal axis to switch from one set to the other. Mechanical or hydraulic cylinders are used for rotation. Gage wheels and rear wheels are automatically repositioned as the plow bottom frame rolls over, unless each set of bottoms has its own wheel (common in mounted plows). Two-way plows eliminate back and dead furrows, creating a more level field for irrigation or drainage. They are particularly useful for terraced fields, contour ploughing, and small irregular-shaped fields. The animal-drawn version of a two-way plow is called a turn-wrest plow.

ADJUSTMENTS OF MOULDBOARD (MB) PLOUGH

Adjustments of a mouldboard plough are essential for optimizing its performance and ensuring efficient soil tillage under varying soil conditions. Proper adjustments allow the operator to control the depth of ploughing, the angle at which the mouldboard interacts with the soil, and the overall effectiveness of soil turning. Below are the main adjustments that can be made on a mouldboard plough:

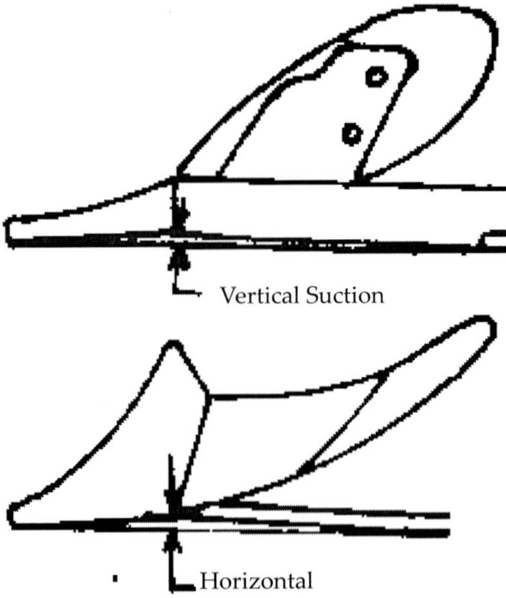

Figure 1.21: View of Adjustment of Mouldboard Plough.

1. **Vertical Suction:** Vertical suction is an adjustment that ensures a moldboard plow enters the soil at the correct depth. It is measured as the maximum distance between the ground surface and the landside when the plow is in its standard working position. Typically, the ideal vertical suction ranges from 3 to 5 millimeters. This distance is measured at the point where the share and landside meet, facilitating proper penetration of the plow into the soil.

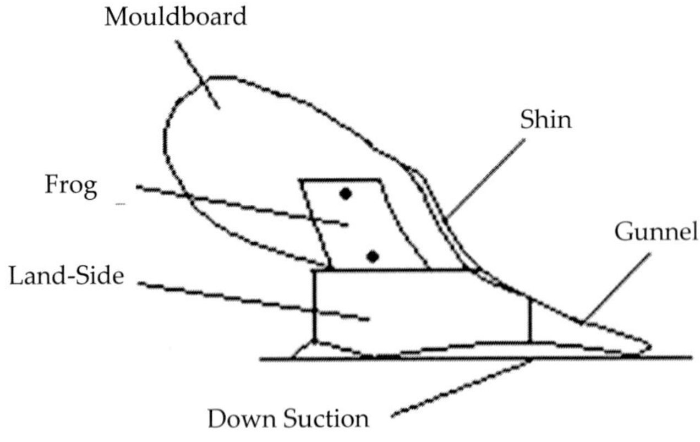

Figure 1.22: Vertical Suction

2. **Horizontal Suction:** Maximum clearance between the landslide and horizontal plane touching point of share at its gunnels side and heel of landslide. It helps the plow to cut the proper width of furrow slice. It is generally 5-13 mm. Angle of cutting edge from direction of travel is 40-45 degrees.

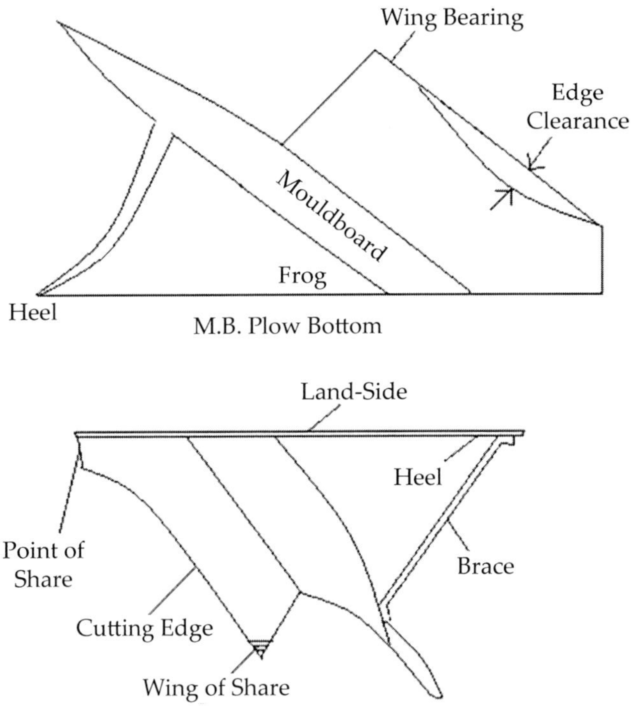

Figure 1.23: Horizontal Suction

3. **Throat Clearance:** Perpendicular distance between point of share and lower position of beam of plow.
4. **Vertical Clevis:** It is a vertical plate with a number of holes at the end of the beam to control the depth and adjust the line of pull.

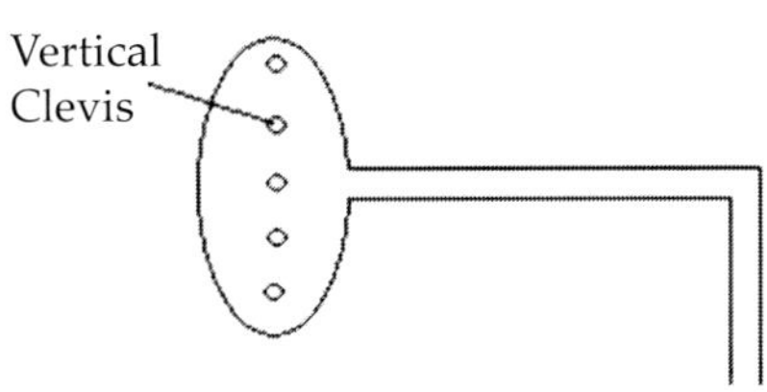

Figure 1.24: Vartical Clevis

5. **Horizontal Clevis:** It is a device to effect lateral adjustment of plow relative to line of draft.
6. **Plow Size:** Perpendicular distance from wing of share to the line joining the point of share and heel of landslide. The size of Mouldboard is expressed by width of cut of soil. Common size of tractor drawn Mouldboard is 30 cm, 35 cm, and 40 cm. Animal drawn Mouldboards are of 13 to 20 cm.
7. **Mouldboard Plow Cross Shaft:** It is used to make proper width adjustment of plow.

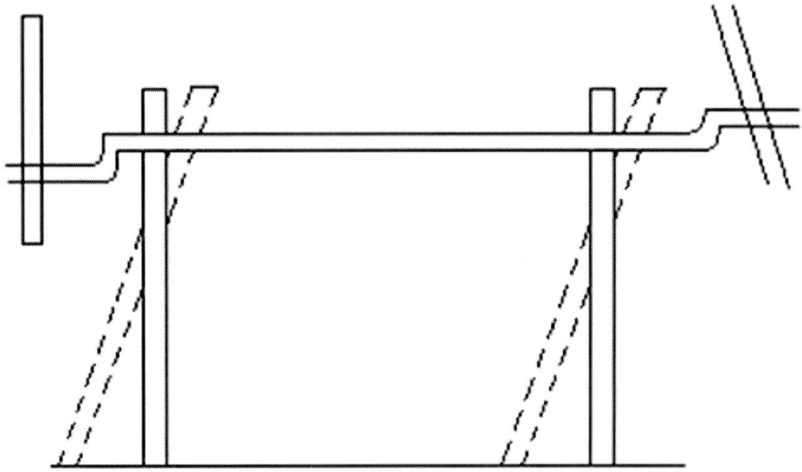

(A) To Increase Forrow Width

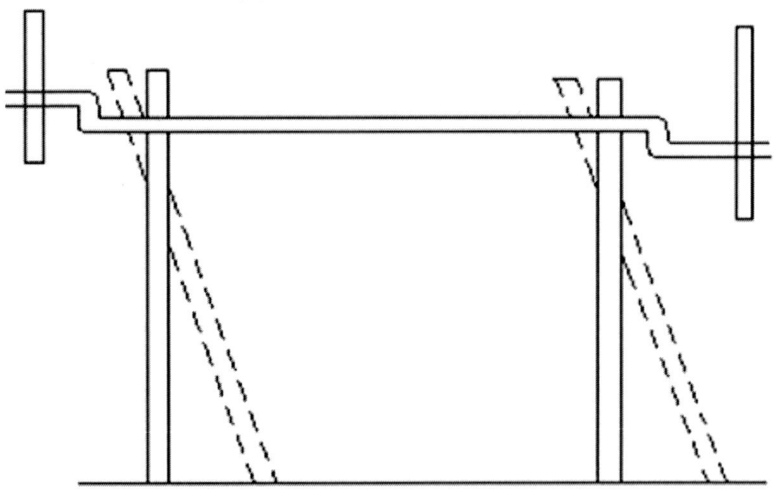

To Decrease the Furrow Width

Figure 1.25: (a) Increase of Furrow Width and (b) Decrease of Furrow Width

8. For tractor drawn plows:

   Coulter adjustments:

   - Vertical adjustments = ½ of depth or 50 mm
   - Horizontal adjustments = 20 mm outside the landslide.
   - Coulter diameter = 30 to 45 cm.

   Jointer adjustments:

   - Depth = 4 to 5 cm
   - Furrow wheel adjustments: 6 to 12.5 mm clearance between rear of landslide back plow bottom and furrow wall.

Working of Moulboard Plow

A moldboard plow features large, curved blades that dig deep into the soil as it is pulled across the land by a tractor. It cuts a row, simultaneously turning and aerating the soil, creating a furrow for planting with a mound of loose, aerated soil on the side. Additionally, it helps incorporate fertilizer that was spread before plowing.

A reversible plow, also called a reversible moldboard plow, can turn the furrow slice to either the right or left side during each round. It has two sets of different bottoms, allowing all furrows to be turned in the same direction by using one bottom for one direction and the other bottom for the return. These two sets of bottoms are designed to be raised or lowered independently or rotated along an axis. The use of a reversible plow prevents disturbance to the land and avoids leaving dead furrows or back furrows in the middle of the field.

Applications

This is employed for primary tillage tasks. It effectively cuts and completely buries trash. It is also used to incorporate green manure crops into the soil, allowing them to decompose and enrich the soil with humus. Additionally, it aids in mixing compost, farmyard manure, or lime into the soil.

Benefits of Mouldboard Plough

- The MB plough is advantageous for marshy regions as it improves soil drainage.
- It ensures complete soil inversion, leading to an even distribution of humus and organic matter on the surface.
- The thorough inversion process also aids in controlling weeds.

DISC PLOUGH

A disc plough is a type of plough that utilizes concave discs to break up and turn the soil. Unlike the traditional mouldboard plough, which uses a curved metal blade to flip the soil, the disc plough employs rotating discs to cut and turn the soil, offering better performance in certain soil types and field conditions. It is particularly effective in handling tougher, compacted soils, stony areas, and fields with crop residues. The disc plough consists of a frame with one or more discs, mounted at an angle to the oil surface, that slice and cut through compacted layers instead of flipping the soil.

Disc ploughs are particularly useful in conditions where mouldboard ploughs may face difficulties, such as in heavy, sticky soils or when working with crop residues. The discs' ability to cut into the soil without fully turning it makes the disc plough well-suited for tough conditions. Disc ploughs come in single-way and double-way variations, depending on the mounting direction and angle of the discs.

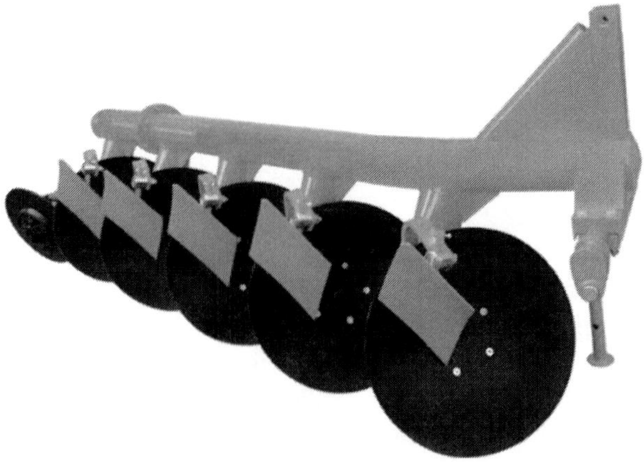

Figure 1.26: Disc Plough

Structural Feature of Disc Plough

Frame

The frame of a disc plough serves as the primary structural support, holding the discs and connecting them to the tractor or draft animals. Typically made of steel, it is built to withstand the forces during ploughing. The frame provides stability to the entire machinery and is often made of cast iron for durability. It can be mounted to a tractor and is responsible for evenly distributing the plough's weight and attaching it to the tractor via the three-point linkage system.

Function: The frame secures the discs in place, either rigidly or with some flexibility, depending on the type of plough. It also ensures proper weight distribution and attachment to the tractor.

Disc

The discs are round, concave plates usually made of hardened steel, typically ranging from 4-6 mm in thickness. These discs rotate as they cut into the soil, with their concave shape aiding in lifting and turning the soil. Discs are the primary functional components of the plough and are attached to a horizontal shaft. The number of discs can vary from 2 to 12, depending on the specific requirements.

Types of Discs

- **Plain Discs:** Smooth discs without any additional features.
- **Notched Discs:** Serrated or notched discs designed for cutting through tough soils and plant residues.
- **Function:** As the plough is pulled, the discs rotate, cutting into the soil and turning it to form furrows. The design of the disc helps break compacted soil and turn it effectively.

Hitch or Linkage

The hitch connects the disc plough to the tractor or draft animal. In modern ploughs, a three-point hitch system is commonly used for easy attachment and detachment.

Function: The hitch allows for raising and lowering the plough during operation, helping adjust the working depth and ensuring the plough stays properly aligned with the tractor.

Beam

The beam is the central component of the frame that supports and holds the discs in place, typically located at the center of the plough.

Function: The beam serves as the main support for the discs. In a single-body disc plough, it maintains the correct angle for soil engagement, and in multi-body ploughs, it connects several disc units.

Horizontal Shaft

The horizontal shaft provides an axis of rotation for the discs, playing a key role in creating the necessary angle for cutting the soil.

Bearings and Spindles

Bearings and spindles are essential components that support and enable the discs to rotate freely. The bearings are located within the frame, while the spindles are directly connected to the discs. Bearings help ensure smooth movement of the blades and prevent the buildup of dirt and debris.Function: These parts allow the discs to rotate freely as the plough moves through the soil. Proper maintenance and lubrication of the bearings are crucial for smooth operation.

Furrow Wheels

Furrow wheels are positioned at the backside of the plough to help regulate the depth of ploughing. They also assist in keeping the plough on track, ensuring a consistent furrow and preventing tilting or instability.

- **Function:** These wheels can be adjusted to achieve different ploughing depths and help maintain proper alignment during operation. Depending on the plough's design and size, some models feature a single furrow wheel, while others may have two or more.

Skids

Some disc ploughs are equipped with skids, small metal plates that glide along the soil surface to guide the plough and prevent the discs from sinking too deeply into the soil.

- **Function:** Skids are especially beneficial in wet or sticky soils, preventing the plough from getting stuck. They also give stability by ensuring the discs are not pushed too deep into the soil.

- **Depth Control Mechanism:** The depth control mechanism is designed to adjust the ploughing depth by altering the angle of the discs or adjusting the frame height.
- **Function:** Depth control can be adjusted either hydraulically or manually, allowing for easy changes during ploughing. In some models, the depth control is linked to the tractor's hydraulic system, facilitating smooth adjustments while in operation.

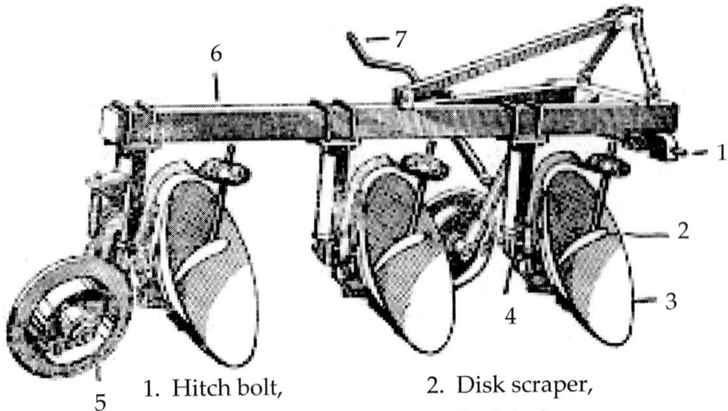

1. Hitch bolt,
2. Disk scraper,
3. Disc,
4. Disk hub,
5. Rear furrow wheel,
6. Beam or Frame, and
7. Lift crank

Figure 1.27: Bearings and Spindles

Varieties of Disc Ploughs

Disc plows are classified into the following main categories:

- Bullock Drawn
- Sulky Type
- Gang Type: Attached to a universal frame mounted on two wheels.
- Tractor Drawn
- Direct Mounted
- Semi-Mounted
- Trailed
- According to the Mounting of Discs:
- Standard Disc Plow
- Vertical Disc Plow (also known as Harrow Plow or Wheat Land Plow)
- Cylinder Plow (also called Disc Tiller or Tiller Disc Plow)

Standard Disc Plow

A Standard Disc Plough is a widely used type of disc plough, designed for general agricultural applications, especially in moderately compact or stony soils.

It is built to handle a variety of soil conditions and provides reliable performance during tillage operations. The standard disc plough typically consists of a frame that supports several concave discs, which are responsible for cutting and turning the soil.These ploughs feature a conventional concave disc configuration, with the discs aligned horizontally and connected to a common shaft. The design drives power through the axial rotation of the blades as they cut and turn the soil. The number of discs typically ranges from 3 to 13.

Standard disc plough types include a three-bottom mounted version, a six-bottom trailed version, and a four-bottom mounted reversible version. The frame holds a series of inclined disc blades, supported by wheels, which are spaced to cut between 18 to 30 cm per disc. The discs are tilted backward at an angle of 15 to 25 degrees from the vertical, and the disc angle is 42 to 45 degrees from the direction of travel.The diameter of the discs typically ranges from 61 to 71 cm. Scrapers are used to help the discs cover trash and prevent soil build-up, especially in sticky soil conditions. Reversible disc ploughs (often mounted or semi-mounted) Consists a mechanism to reverse the disc angle at the ends of the field, allowing for one-way plowing. These ploughs are commonly used in hard and dry soils, where the discs are driven into the ground by gravity rather than suction.

The total weight of a disc plough ranges from 200 to 600 kg. In mounted types, the tractor's wheels bear the side thrust, with a rear wheel sometimes added to reduce the load. In trailed types, the furrow wheel handles the side thrust. The discs are made from heat-treated steel with a thickness of 5 mm to 10 mm.

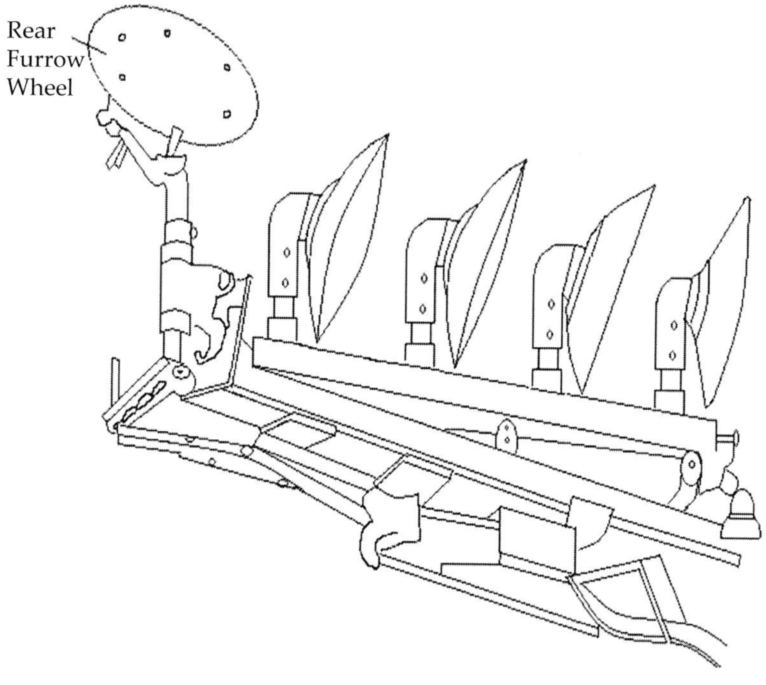

Figure 1.28: Standard Disc Plow

Application

- The horizontal disc arrangement is ideal for cutting the upper layers of soil and inverting it.
- The Standard Disc Plough is highly effective in managing residues and controlling weeds.
- It is widely used by farmers for mixing the soil.
- Typically, it is employed during primary tillage operations, particularly in fields with compacted and heavy soils.

Merits

- It offers superior ploughing performance during primary tillage compared to other plough types.
- It effectively buries weeds and harmful residues, making it useful for weed control.
- It is capable of quickly breaking up large soil clods.
- Standard disc ploughs can adapt to various soil conditions.

Demerits

- It is not designed for deep soil penetration and is less effective during secondary tillage.
- Repeated use of the standard disc plough increases the risk of soil erosion.
- It requires more power for tillage compared to a vertical disc plough.
- Consistent use can potentially damage the humus layer of the soil.

Vertical Disc Plow

Vertical disc plow is also known as the one-way disc plow, harrow plow, and wheat land plow full stap. The vertical disc plough features concave blades arranged in a V-shaped configuration. The discs maintain a flat angle with the shaft but form an angle with the soil, allowing them to penetrate deeper compared to a standard disc plough. Similar to the pull-type and standard disc plough, the discs on a vertical disc plough are evenly spaced along a common axle or gang bolt and are held together with spacer spools, allowing the entire gang to rotate as a unit, much like a disc harrow.

It is positioned between the standard disc plough and the disc harrow in terms of its soil working action. This type of plough is generally used in flat areas where shallow plowing and the mixing of stubble with soil are needed. The diameter of the discs ranges from 51 cm to 61 cm, with a spacing of 20 to 25 cm along the gang bolt. The width of the cut per disc depends on the angle between the gang axis and the direction of travel, with disc angles varying from 35 to 60 degrees, and the most common angle being 40 to 45 degrees. The width of the cut from various sizes of vertical disc ploughs ranges from 2 to 6 meters, and the weight of the plough ranges from 50 to 100 kg.

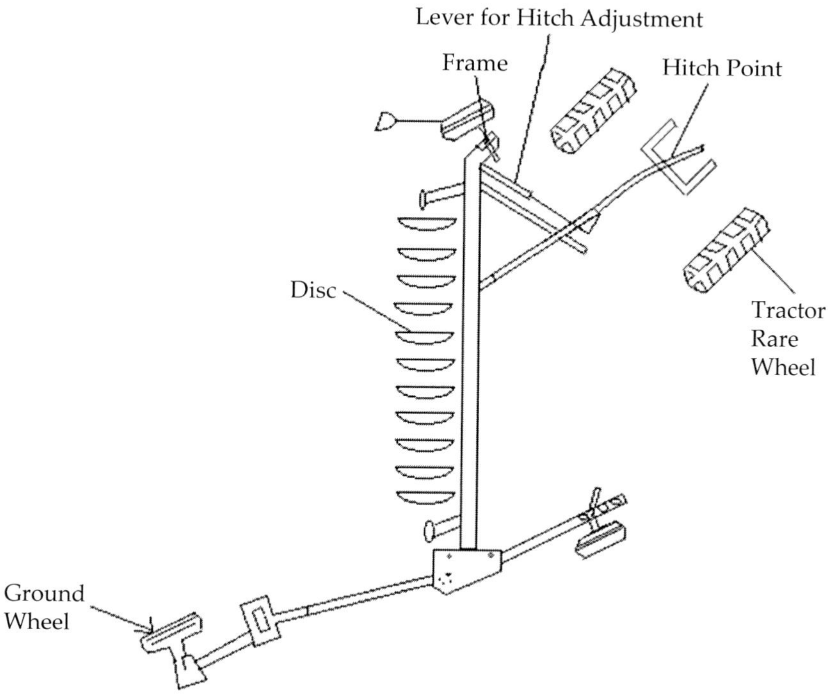

Figure 1.29: Vertical Disc Plough

Applications

- It is suitable for both primary and secondary tillage activities.
- It facilitates deep soil inversion, turning over the humus layer.
- It effective in preparing hard and marshy soils for cultivation.
- It helps in preventing soil erosion.
- It is useful for creating furrows for irrigation and seedbeds.

Merits

- It provides improved soil inversion during secondary tillage activities.
- It is more efficient than traditional disc ploughs.
- Its unique soil inversion technique helps in preventing soil erosion.
- It is can also be used for removing weeds and other unwanted plants from the soil.

Demerits

- Continuous use of vertical disc ploughs can harm the upper soil layer.
- High-speed cutting is not feasible due to the vertical alignment of the discs.
- Can leave the field drier and cloddier compared to a standard disc plough.

Table 1.1: Comparison between Standard Disc Plow and Vertical Disc Plow

S. No.	Comparison	Standard Disc Plough	Vertical Disc Plough
1.	Mounting of disc	Individual axis	Common axis
2.	Number of discs	1-6	5-24
3.	Sizes of discs	60-75 cm	40-60 cm
4.	Spacing	18-30 cm	18-22 cm
5.	Concavity	More	Less
6.	Depth of cut	30-40 cm	8-10 cm
7.	Angle to direction of travel	42-45 degree	35-60 degree
8.	Tilt angle	15-25 degree	0 degree
9.	Weight/disc	200-600 kg	50-100 kg
10.	Draft requirement	High	Low

In standard disc plow provisions are made for adjustments of disc angle and tilt angle.

Disc Angle

The disc angle refers to the angle between the cutting edge of a disc and the direction of travel. It typically ranges from 42 to 45 degrees.

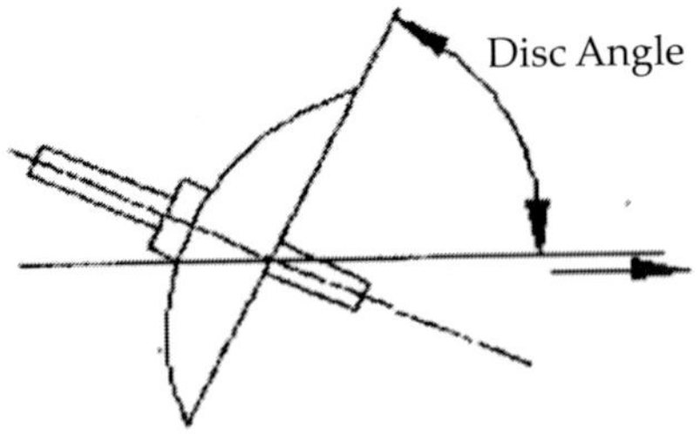

Figure 1.30: Disc Angle

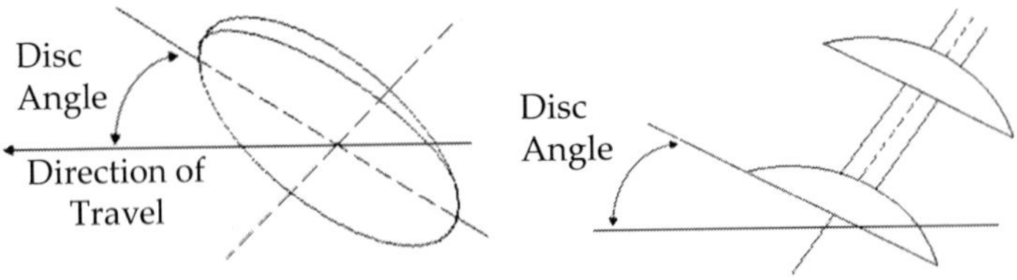

Figure 1.31: Schematic of Disk Angle

The angle of the disc provides a primary guide for surface penetration. This is one of the main parameters that you can use to improve productivity.

However, adjusting the angle of the disc incorrectly can cause some unwanted problems. The angle refers to the inclination that the discs make with the ground. It provides direction and controls the movement of the blades in the direction of travel.

If the angle becomes larger, it can aggressively destroy the soil surface and damage the humus content. On the contrary, a smaller disc angle can reduce the penetration depth and result in shallow tillage operations. It is therefore necessary to perfectly adjust the disc angle for better efficiency.

Tilt Angle

It is the angle at which the plane of cutting edge of disc is inclined to vertical line. Tilt angle varies from 15 to 25 degree.

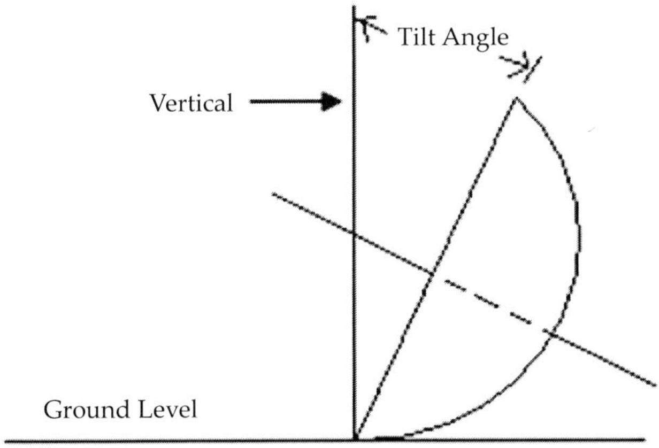

Figure 1.32: Tilt Angle

The disc angle is used for the horizontal movement of the disc and on the other hand the tilt angle ensures the vertical movement of the discs. The tilt angles determine how effectively the discs can return to the ground after penetration.

A steeper angle can cause excessive soil disturbance and result in increased pavement damage. It can also lead to unnecessary fuel consumption and loss of power. Conversely, if the tilt angle is too large, the disc cannot create the desired reversal.

It can be used for limited weeding and rough abrasion. The ideal slope is considered between 15 and 25 degrees. Correct adjustment of the slope helps us improve the inversion and leveling of the ground.

- Adjustment of tilt angle
- Shifting standard on the frame
- By adjusting blocks
- Changing grooves in bearing brackets

Components of a Disc Plough

- **Disc blade:** Disc type blades are mounted for cutting of soil. Number of blades and diameter determine plough capacity. Concavity affects disc angle and soil turning. Shallow concavity depends on diameter of discs. Depth of cut depends on diameter of discs. About 1/3rd of blade diameter is the limit for depth. Width of cut depends on diameter of blade. Width of cut is normally 0.4 times of diameter of disc blade.

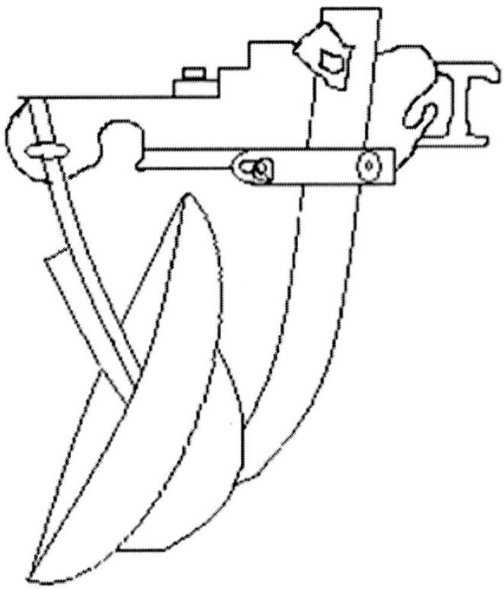

Figure 1.33: Dise Blade

- **Standard:** Connects disc bearing and plough frame. Sometimes beam is bent for disc attachment for reducing the cost.

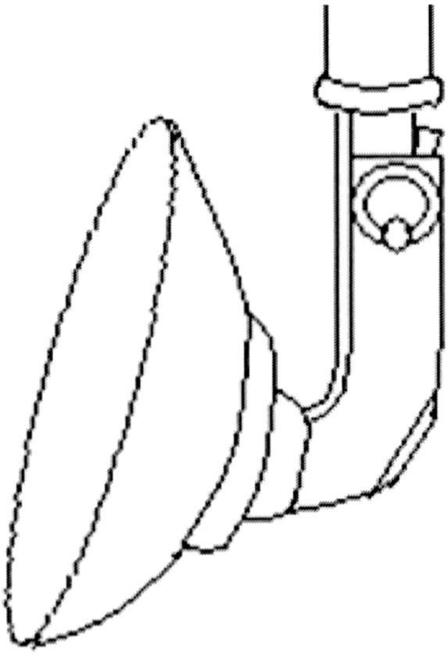

Figure 1.34: Standard Blade

- **Plough Frame:** Standards are attached to the plough frame. It has provision for disc angle adjustment, adding or removing standard and discs.
- **Rear Furrow Wheel:** To stabilize the plough and take side thrust, rear furrow wheel is fixed at end.

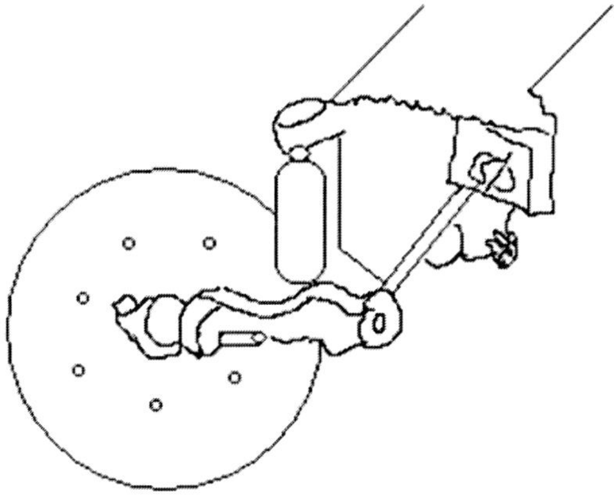

Figure 1.35: Rear Furrow Whell

- **Bearings:** Disc blades are at an angle to the direction of travel so both radial and thrust forces are present. Radial forces push against an axle at right angle while thrust forces push along the axis. That is why taper roller bearings are used.

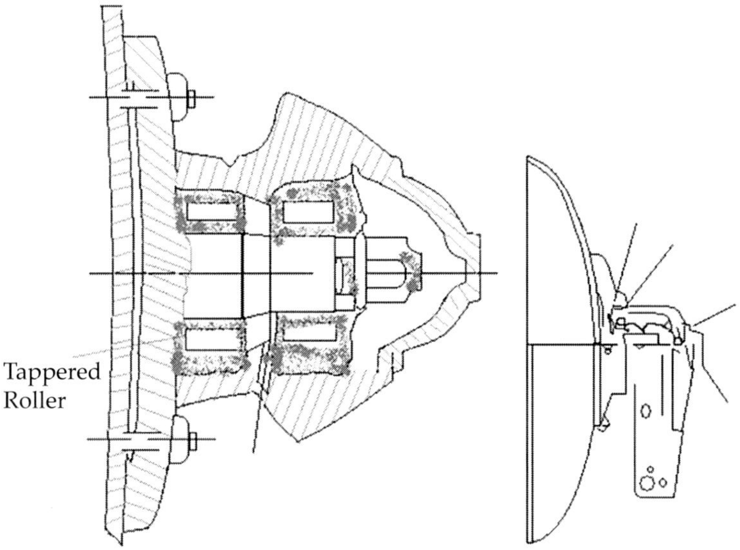

Figure 1.36: Bearings

- **Scraper:** Device to remove soil that tends to stick to working surface of a disc, for great pulverization to furrow slice, invert the furrow slice and cover the trash better.

Types of Scrapers

Scrapers are tools used for various purposes such as cleaning, shaping, or smoothing surfaces. There are different types of scrapers, each designed for specific tasks. Below are some common types of scrapers:

- **Hand Scrapers:** These are manually operated tools used for smoothing, cleaning, or scraping surfaces. They come in various shapes and sizes, including flat, curved, or triangular blades. Thses are Common in woodworking and metalworking.
- **Paint Scrapers:** These scrapers have a flat, wide blade and are designed for removing paint, wallpaper, or adhesives from surfaces like wood, metal, or walls.
- **Carpet Scrapers:** These are used to remove carpet adhesive or glue from floors. They often have a long handle and a sharp, flat blade to scrape off residues.
- **Ice Scrapers:** These are typically used in cold climates, these scrapers have a flat blade to remove ice or frost from car windshields and windows.
- **Floor Scrapers:** These are used for removing tile adhesive, linoleum, or old flooring materials. They come with long handles and wide blades for effective surface scraping.
- **Wire Scrapers:** These tools are used for cleaning and removing debris from surfaces, often used in welding or heavy-duty industrial applications.
- **CNC Scrapers:** In machining, these scrapers are used to smooth and finish metal or other materials to a precise level.
- **Surgical Scrapers:** Theses are used in medical procedures for scraping tissue or bones during surgery. They are small, precise, and come with sharp, curved edges.
- **Dental Scrapers:** These are specialized tools used by dentists for cleaning plaque or tartar from teeth.
- **Ice Cream Scrapers:** These have a curved edge for scraping and serving ice cream or frozen desserts from containers.

Each scraper type is optimized for its specific application, whether in construction, medical procedures, or everyday tasks.

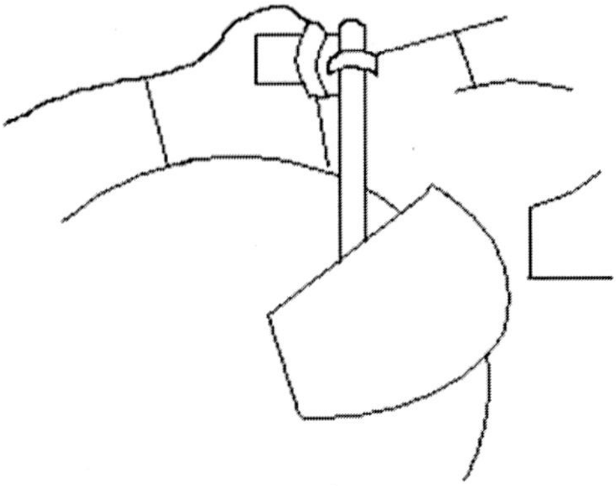

Figure 1.37: Disc Scraper

- **Concavity:** It is the depth measured at the center of disc by placing its concave side on flat surface.

Draft of Disc Plow

Soil type and moisture content are the primary external factors influencing the draft of any plough. In very hard soil, additional weight may be required on the wheel frame to help the plough penetrate the ground. The condition of the bearing housing also plays a role in the draft; it is important to maintain smooth bearings by applying grease when necessary. The draft is further influenced by the depth and width of the cut per bottom of the plough. Speed also affects the draft, with doubling the speed increasing the draft by approximately 20%-25%.

Adjustments to a disc plow can affect the depth, width, and soil pulverization in the following ways: i) Increasing the disc angle improves penetration. ii) Reducing the tilt angle on a standard disc plow enhances penetration. iii) Adding extra weight to the plough also improves penetration.

The depth of penetration typically ranges from 3″ to 15″. For optimal penetration, discs should be kept sharp.

Width of Cut

This can be adjusted by changing the angle between the land wheel axle and the frame. It can also be improved by increasing the disc angle and reducing the tilt angle.

To Compute Radius of Curvature of Discs:

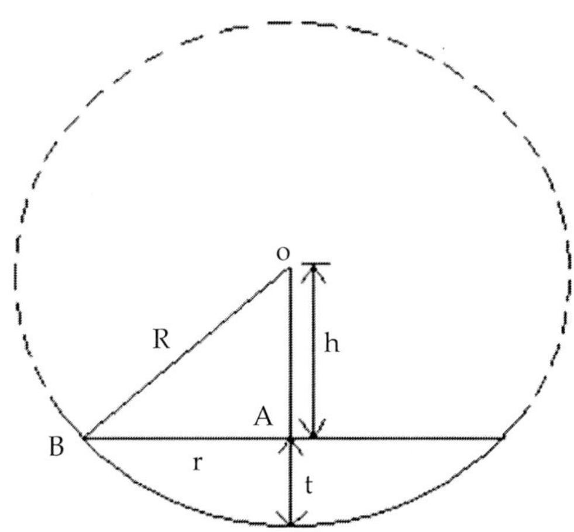

R – Radius of Curvature

r – Radius of disc

t - Depth or concavity

OAB

$R^2 = h^2+r^2$

Also, h = R-t

$$h^2 = (R-t)^2$$
$$h^2 = R^2 - 2Rt + t^2$$
$$R^2 = R^2 - 2Rt + t^2 + r^2$$
$$2Rt = t^2 + r^2$$
$$2R = \frac{t^2 + r^2}{t}$$
$$2R = t + \frac{r^2}{t}$$
$$2R = t + \frac{(D/2)^2}{t}$$
$$2R = t + \frac{D^2}{4t}$$

Thus, we get

$$R = \frac{t}{2} + \frac{D^2}{8t}$$

Table 1.2: Trouble Shooting

	Defect	Reasons	Remedy
1.	Low Penetration	Blunt disc	Sharpen the edge
		Plow too light	Put additional load
		More tilt angle	Set tilt angle
2	Heavy draft	Blunt disc	Sharpen the edge
		Furrow too wide	Reduce tilt angle
3.	Excessive Side Draft	Improper hitching	Hitch properly
4.	Less Scouring	Defective Scraper	Adjust the Scraper
5.	Uneven Furrows	Disc angle not uniform	Set disc angle
		Loose bearings	Set bearings
		Hitching defective	Set hitching

Advantages of Using a Disc Plough

- It is capable of digging deep into the soil, making it ideal for stony and rough terrains.
- The plough offers adjustable cutting depths, allowing farmers to tailor tillage activities according to their needs.

- Its robust construction enables it to work effectively on both flat and uneven surfaces.
- It helps reduce soil erosion and supports sustainable land management practices.
- Disc ploughs improve soil aeration while minimizing clogging.

Adjustments for Optimizing Disc Plough Performance

- **Depth Adjustment:** Adjusting the depth is crucial for the performance of a disc plough, as it directly influences soil inversion and weed control. Proper depth ensures healthy crop development, while incorrect depth settings can disrupt soil aeration, nutrient distribution, and hinder root growth. Shallow depth may prevent proper turnover, while excessive depth can damage the soil's nutrient-rich layers and lead to erosion. Adjusting the depth based on soil conditions is essential.
- **Angle Adjustment:** The angle of the disc blades determines the cutting and turning behavior of the plough. Adjusting the angle affects the aggressiveness of the cutting action—an increased angle leads to more aggressive cutting, while a decreased angle softens the action. Angle adjustments are especially important when ploughing stony soils or varying vegetation densities, offering flexibility for different soil conditions.
- **Gang Adjustment:** Modern disc ploughs often feature multiple disc gangs that can be adjusted individually. This allows for precise customization of the ploughing operation, enabling farmers to create ridges, furrows, or level surfaces with accuracy. Fine-tuning the gang settings improves efficiency, deepens penetration, and helps conserve power, ensuring the plough digs deep with fewer blades.
- **Hitch Adjustment:** Proper hitch adjustment ensures that the disc plough is properly aligned and balanced with the tractor. This alignment increases the plough's performance and reduces wear on both the plough and the tractor, ensuring long-term reliability. Hitch adjustment also strengthens the disc movement, improving soil inversion during operation.
- **Blade Maintenance:** Regular blade maintenance is crucial for consistent performance. Keeping blades sharp, replacing worn-out ones, and lubricating moving parts prevents rust and corrosion, extending the lifespan of the equipment and ensuring efficient soil preparation.
- **Leveling the Plough:** The plough's level is controlled by the tractor's top link. If the rear of the plough beam is higher than the front, the top link should be lengthened; if it's lower, the top link should be shortened. Lateral leveling is adjusted by modifying the length of the right lower link. These adjustments should be made before starting the operation.
- **Tightening the Bearing:** Bearings should be kept properly tightened. Tighten the castle nuts until the disc firmly binds to the hub.

- **Scraper Adjustments:** Scrapers should be set low enough to catch and turn the furrow slice before it falls away from the disc. For deeper ploughing, set the scraper slightly higher, while for sticky soils, adjust the scraper closer to the disc. Research indicates that the mouldboard-type scraper is most effective, but a hoe-type scraper is better for sticky soils.

Figure 1.38: Disc Plough Having Scrapers.

Applications

Disc plough is used for primary tillage and is especially useful in hard and dry, trashy, stony or stumpy land conditions and in soil where scoring is a major problem

Merits of Using a Disc Plough

- A disc plough can effectively penetrate hard and dry soil, where a mouldboard plough may struggle.
- It performs well in sticky soils where a mouldboard plough is less effective.
- It is ideal for deep ploughing.
- Farmers can work safely in stony and stumpy soils without the risk of damage.
- It continues to perform effectively even when a significant portion of the disc is worn down in tough soil.
- It works efficiently in loose soils, with minimal clogging.

Demerits of Using a Disc Plough

- A disc plough is less effective at covering surface debris and weeds compared to a mouldboard plough.
- It leaves the soil in a rough, cloddy condition, unlike the smoother finish produced by a mouldboard plough.
- A disc plough is heavier than a mouldboard plough with the same capacity, as its penetration is primarily influenced by its weight, whereas a mouldboard plough is driven into the soil by suction.

Table 1.3: Competitive Analysis Between Disc Plough and Mouldboard Plough

Criteria	Disc Plough	Mouldboard Plough
Soil Penetration	Effective in hard, dry, and sticky soils.	Best for softer soils; struggles in tough, dry soils.
Performance in Stony/Uneven Terrain	Performs well in stony and stumpy areas.	May face damage or difficulty in stony or uneven soils.
Depth of Ploughing	Ideal for deep ploughing.	Suitable for moderate depth ploughing.
Suitability for Surface Trash	Less effective at covering surface debris and weeds.	More effective at covering trash and weeds.
Soil Finish	Leaves soil rough and cloddy.	Leaves soil smooth and fine.
Weight	Heavier; relies on weight for penetration.	Lighter; relies on suction for penetration.
Clogging	Minimal clogging, especially in loose soils.	More prone to clogging in wet or sticky soils.
Maintenance	Requires regular blade maintenance.	Generally requires less maintenance than disc plough.
Safety in Rough Conditions	Safe to use without risk of damage in rough terrain.	May risk damage in stony or rough conditions.
Efficiency in Loose Soil	Performs well with minimal clogging.	Effective in loose soil but more prone to clogging.

This table highlights key differences and advantages between the Disc Plough and the Mouldboard Plough, helping to choose the right tool based on soil conditions and ploughing requirements.

DISC PLOUGH MAINTENANCE

Maintenance Instructions

- Operating a disc plough on stony terrain requires increased maintenance. Follow these guidelines to ensure optimal performance:
- After the first two hours of use, tighten all nuts and bolts if the disc plough is new.
- Check the plough adjustments if the steering becomes stiff.
- After 300 hours of use, open the disc plough hub, clean it with diesel oil, apply new grease, and replace the seal.
- Replace the grease nipple if soil has entered it.

- Grease all lubrication points with a grease gun every fifty hours and tighten all nuts and bolts.
- Regularly inspect the scrapper adjustments.
- Replace the discs when their diameter drops below 22″ (550 mm) for effective ploughing.
- Sharpen the discs if the blades are dull, as blunt blades can significantly increase draft.
- When the disc diameter reduces to 24″ (600 mm), adjust the hub by loosening the mounting bolts.
- Continuously check for loose nuts and bolts.
- Lubricate the bearings according to the manual instructions.
- To prevent rust, coat the disc blades with used oil during the off-season.

Storage of the Machine After Work

- Wash the disc plough after use.
- Replace any worn-out nuts and bolts.
- If the disc plough will not be used for an extended period, clean it and apply a layer of used oil to prevent rust.
- These steps will help extend the life of your disc plough.

Lubrication

- Ensure that high-quality grease is used in the bearing housings, coulter hub, and bushes.
- Lift the disc plough on every turn.
- Before ploughing check all nuts and bolts of the disc plough.
- Never plough on stony soil.
- Tractor should be in high first gear.
- If the soil is hard then plough the field at least twice.
- Make sure that the shocker spring is tight.
- Warnings for the driver
- Be vigilant about the tree roots and stones.
- Keep proper distance from disc plough when disc plough is in working.
- Lift the plough before approaching the road.

Safety Considerations

- To ensure the operator's safety, always wear appropriate clothing and footwear during operations.
- Never allow passengers on the tractor or implement unless an additional seat is provided.
- Exercise caution when operating on steep gradients to prevent tipping.
- Avoid transporting the implement on rough roads at night.

- When operating, steer clear of sharp turns that could cause contact with the implement.
- Never stand between the disc plough and the tractor when the plough is ready for use.
- Ensure the three-point linkage is correctly fitted and secured with a lynchpin.
- If the scrapper touches the discs, loosen the scrapper bolts and adjust it properly.
- Never turn the tractor left or right while the plough is engaged in the soil.
- Never reverse the tractor when the plough is engaged in the soil.

CHISEL PLOUGH

The chisel plough is a widely used tool for deep tillage with minimal soil disturbance. Its primary purpose is to loosen and aerate the soil while keeping crop residue on the surface. This plough helps alleviate compaction and break through ploughpan and hardpan layers. Unlike other ploughs, the chisel plough does not invert or turn the soil. This feature has made it especially beneficial for no-till and low-till farming practices, which aim to preserve organic matter and farming residues on the soil surface throughout the year, thus enhancing erosion prevention.

Figure 1.39: Chisel Plough

Chisel plough is a unique agricultural implement that includes several narrow and tapered shanks. The shanks move vertically to penetrate the soil and help in inversion. Chisel plough is very effective for stony and marshy soil conditions.

Main Components of Chisel Plough

- **Shanks:** Narrow, pointed metal bars that penetrate the soil.
- **Frame:** The main structure that holds the shanks in place.
- **Depth Control Wheels:** Help to maintain consistent ploughing depth.

Benefits of Using Chisel Plough

- Minimizes soil erosion by not turning over the soil.
- Preserves soil moisture and structure.
- Improves soil aeration and root penetration.

Limitations of Using Chisel Plough

- Less productive in comparison to mouldboard and disc plough.
- Complex in use.
- Provide minimal weed control.

SUBSOILER PLOUGH

Subsoilers are specifically designed for deep tillage tasks. They are particularly effective in mountainous and rocky regions. A subsoiler plough can reach depths of up to 60 cm in a single pass.

Figure 1.40: Subsoiler Plough

Key Components of Subsoiler Plough

- **Shanks:** Deep-penetrating metal bars.
- **Frame:** The structure that holds the shanks.
- **Gauge Wheels:** Help to control the depth of penetration.

Benefits of Using Subspoiler Plough

- Chisel ploughing loosen and aerate the soil
- Reduces the effect of compaction
- Can help break up hardpan
- Reduces soil erosion low till farming conditions
- Prevents formation of surface crusts, which helps water to infiltrate the soil
- Chisel ploughing severe weed roots below the surface.

Limitations of Using Subspoiler Plough

- High risk of soil erosion in sloped areas.
- It costs high energy.
- Application is limited to some kinds of soils.

ROTARY PLOUGH

Rotary ploughs, also known as rotavators, are traditional ploughing machines used for primary tillage. They feature rotating blades that break up and pulverize the soil. Rotary ploughs are particularly effective for cultivating seasonal crops and working in marshy soils.

Figure 1.41: Rotary Plough

Key Components of Rotary Plough

- **Rotary blades/tines:** The rotating components that break up and pulverize the soil.
- **Rotor shaft:** The shaft that holds the blades/tines in place.
- **Gearbox:** Transfers power from the tractor to the rotary blades.

Benefits of Using Rotary Plough

- Creates a fine, well-prepared seedbed.
- Evenly incorporates crop residues and organic matter into the soil.
- Ideal for small-scale and intensive farming operations.

Limitations of Using Rotary Plough

- Less effective in controlling weed..
- Less useful during secondary tillage activities.
- Not efficient for large-scale farming.

RIDGING PLOUGH

The ridging plough is an effective as well as efficient tool for creating ridges and furrows during seedbed preparation. However, with the introduction of hydraulic reversible ploughs, ridging ploughs are now scarcely used by farms, though they remain an excellent option for field preparation.

Figure 1.42: Ridging Plough

Main Components of Ridging Plough

- **Mouldboards:** Shaped specifically to form ridges.
- **Frame:** Holds the mouldboards securely.
- **Gauge Wheels:** Control the depth and width of the ridges.

Benefits of Using Ridging Plough

- Enhances drainage and prevents waterlogging.
- Improves soil aeration, supporting better root development.
- Eases irrigation and weed control in row crops.

Limitations of Using Ridging Plough

- Creates raised planting rows.
- Can increase erosion by altering runoff patterns.
- Best suited for tuber and root crops.

HYDRAULIC REVERSIBLE PLOUGH

Hydraulic reversible ploughs are increasingly dominating the agricultural sector due to their superior productivity and efficiency compared to traditional ploughs.

Figure 1.43: Hydraulic Reversible Plough

Main Components of Hydraulic Reversible Plough

- **Reversible mouldboard:** Can be flipped to operate in both directions.
- **Hydraulic system:** Facilitates easy reversal of the mouldboard.
- **Frame:** Supports both the mouldboard and hydraulic system.

Benefits of Using Hydraulic Reversible Plough

- Boosts efficiency by minimizing the need for headland turns.
- Ensures consistent tillage depth and soil structure.
- Adaptable to various soil types and conditions.

Limitations of Using Hydraulic Reversible Plough

- Requires specialized knowledge for maintenance and repairs.
- Higher initial cost and ongoing maintenance expenses.

Final Thoughts

Primary tillage plays a vital role in the cultivation process, and the choice of tillage equipment significantly affects soil health and crop yields. Mouldboard ploughs and disc ploughs are essential tools in primary tillage, each suited to specific soil conditions and farming practices. Familiarity with the design, adjustments, and operational principles of these ploughs is key to effective soil management and maximizing crop productivity.

2

Secondary Tillage Equipment

Secondary tillage equipment refers to agricultural machinery used after primary tillage (such as plowing) to refine the soil for planting. It helps in breaking down clods, incorporating organic matter, controlling weeds, improving soil aeration, and preparing the seedbed. This step is essential for creating ideal soil conditions for seed germination and root development.

Secondary tillage operations are lighter, finer tasks performed after primary tillage to establish the proper soil texture for planting. These operations cause minimal soil inversion or shifting, and they consume less power per unit area compared to primary tillage. The tools used in secondary tillage include various types of harrows, rollers, pulverizers, rotary tillers, tools for mulching and fallowing, and cage wheels.

Key objectives of secondary tillage is:

- To enhance the seedbed by further breaking down the soil.
- To eliminate grasses and weed seeds from the field.
- To chop crop residues and incorporate them into the topsoil.
- To break large clods and create a smooth, level field surface.

TYPES OF SECONDARY / MINIMUM TILLAGE EQUIPMENT

Below mentioned are some of the main types of secondary tillage implements:

CULTIVATOR

A cultivator is an agricultural tool used for secondary tillage, which involves preparing the soil for planting after primary tillage has been done.

HARROW

A harrow is a farming implement used for shallow tillage. After plowing, a harrow is employed to break up soil clumps and smooth the surface, preparing it for seed planting. It helps improve soil texture and create optimal conditions for seed growth. Coarser harrowing is sometimes performed to eliminate weeds. The Kelly Tillage Disc Harrows are known for their durability and require minimal servicing or spare parts.

A harrow is a secondary tillage tool that serves various functions in crop cultivation, including:

- Cutting vegetable matter like corn stalks, cotton stalks, and weeds before plowing, mixing it into the soil for conservation.
- Pulverizing the topsoil to improve the connection between furrow slices and the soil, eliminating air gaps.
- Pulverizing the soil after plowing to create a suitable seedbed.
- Cultivating crops.
- Used in summer fallowing.
- Covering seeds after sowing.

There are different types of harrows, such as disc harrows, spike-tooth harrows, spring-tooth harrows, rotary cross-harrows, soil surgeons, triangular harrows, acme harrows, blade harrows, and reciprocating power harrows.

Major Components of a Harrow

The harrow consists of several key components:

- **Disc:** Revolving steel plates with a circular, concave shape that cut into the soil, made from high-quality steel, and typically range from 35 to 70 cm in size.
- **Gang Angle:** The angle between the axes of the gang.
- **Gang:** An assembly of concave discs.
- **Gang Control Lever:** A lever for adjusting the gang's angle.
- **Gang Axle (Arbor Axle):** The axle on which the set of discs is mounted.
- **Scraper:** Prevents clogging of the machine.
- **Transport Wheel:** Provides mobility for the harrow when moving between fields.
- **Bearing:** Essential for counteracting end thrust.
- **Weight Box:** Adds extra weight to the harrow.
- **Spool/Spacer:** A flanged tube fitted onto the gang axle.
- **Functions of Harrows:** Harrows are used to break up large clods, level the soil, and incorporate organic materials such as crop residues and fertilizers.

Various Types of Harrows

- **Disc Harrow:** Features rotating discs that cut into the soil, effective for both dry and wet conditions, particularly in breaking up tough soil and residues.
- **Tine Harrow:** Uses flexible, spring-loaded tines to stir and level the soil, suitable for lighter, less compacted soils, often used in orchards or vineyards.
- **Chain Harrow:** Consists of chains that drag over the soil, breaking up clods and smoothing the surface, primarily used for fine leveling or finishing work.

Disc Harrow

A disc harrow is used for harrowing tasks, consisting of rotating steel discs mounted on a common shaft. There are two primary types of disc harrows, classified by the arrangement of the discs.

Single Action

This type of disc harrow has two gangs positioned consecutively, with each gang displacing soil in opposite directions. The discs are arranged so that the right-side gang pushes soil to the right, while the left-side gang pushes soil to the left. The gangs are placed end-to-end at an angle greater than 90°, with the discs on each gang oriented in opposite directions to ensure the soil is thrown in opposite directions.

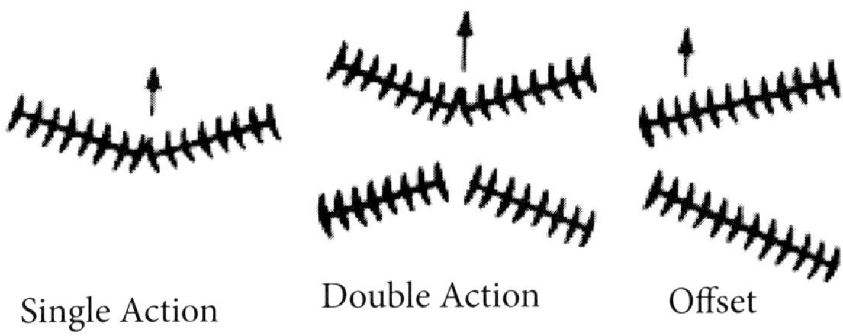

Figure 2.1: Types of Disc Harrow

Double Action

The double action disc harrow features two or more gangs, with one or two gangs following directly behind another set. The gangs are arranged so that the front and back gangs throw the soil in opposite directions. It is often referred to as a tandem harrow, this design involves the front gangs throwing soil outward in one direction, while the rear gangs throw the soil outward in the opposite direction. As a result, the entire field is worked twice during each pass.

Figure 2.2: Double Action

Off-Set Disc Harrow

The offset disk harrow consists of a right-hand gang (throwing soil to the right) and a left-hand gang (throwing soil to the left), working together in tandem. It is called an "offset" harrow because it can be positioned to one side of the tractor. By adjusting the hitch, the harrow can operate to either the left or right of the tractor. This design allows for operation in areas with obstacles, such as under tree branches in orchards.

Key Parts of Disc Harrow

A disc harrow consists of a series of concave cutting metal discs, or blades, mounted on a frame. The primary parts of a disc harrow are as follows:

- **Disc:** Circular concave cutting blades made from steel, available in two types: smooth edge/plain edge discs and notched edge/cut-away or scalloped edge discs.
- **Frame:** The structure that holds and supports the discs.
- **Arbor Bolt:** A heavy steel shaft that mounts the disc onto the harrow.
- **Spool or Spacer:** A flanged tube mounted on the arbor bolt between each pair of discs to prevent lateral movement and maintain the correct spacing between them.
- **Bumper:** A heavy iron plate located at the end of each gang to protect the discs from collisions with adjacent discs.
- **Bearings:** Facilitate the rotation of the disc gang.
- **Scraper:** Prevents clogging by removing soil from the discs, keeping the concave side of each disc clean.
- **Transport Wheel:** Some disc harrows are equipped with wheels for easier transportation.
- **Disc Gang:** The assembly of discs mounted on a common arbor bolt along with the spacer, bearing, and bumper.
- **Weight Box:** Positioned on the main frame to add extra weight, which helps increase the penetration of the discs into the soil.

Applications and Advantages of Disc Harrow

Disc harrows offer several benefits and are used for multiple purposes in agriculture. Below mentioned are some key uses of Disc Harrow:

- **Soil Preparation:** Disc harrows break up and loosen the soil, creating a suitable seed bed for planting. They can cut through and break compacted soil, which in trun which in turn improves water penetration, air circulation, and root growth.
- **Weed Control:** They help control weeds by cutting them into smaller pieces and burying them in the soil, reducing new weed growth and minimizes the need for herbicides.
- **Residue Management:** Disc harrows incorporate crop residues into the soil after harvesting, adding organic matter and improving soil fertility and structure.

- **Improved Crop Yields:** Proper soil preparation with a disc harrow promotes optimal conditions for seed germination and crop growth, leading to better yields.
- **Cost-Effective:** Disc harrows are relatively affordable compared to other farm equipment and can perform a variety of tasks, making them a smart investment for farmers.
- **Time-Saving:** They cover large areas quickly, reducing time and labor essential for field preparation.
- **Versatile:** Disc harrows can be used on various types of and adjusted for different depths and angles, which offers great flexibility for farmers.

Hence, disc harrows are essential tools in modern agriculture, helping farmers prepare their fields for planting, manage weeds, and incorporate crop residues. Their cost-effectiveness, versatility, and time-saving capabilities make them a valuable investment.

Things to Consider When Selecting a Disc Harrow

When selecting a disc harrow following factors should be considered:

- **Soil Type:** Different soil types require different models. For example, heavy clay soils may need a more robust, large-blade harrow.
- **Field Size:** A larger field requires a bigger harrow to cover more ground in less time.
- **Tractor Horsepower:** Ensure the disc harrow is compatible with your tractor's horsepower before purchasing.
- **Number of Discs:** More discs allow for faster and deeper soil penetration, but they also come at a higher cost.
- **Blade Spacing:** The spacing between blades determines the depth of penetration and level of soil disturbance. Wider spacing is better for shallow tilling, while narrower spacing is ideal for deeper tillage.
- **Blade Size:** Larger blades offer deeper soil penetration, while smaller blades create finer tilth.
- **Bearing Type:** The durability and maintenance requirements of bearings depend on the type. Sealed bearings are low-maintenance but may need replacement if damaged, while greasable bearings are more durable but require regular upkeep.
- **Transportability:** Choose a model that can easily be hitched to your tractor for easy transport between fields.
- **Price:** Select a model that fits within your budget while meeting your specific needs.

Tine Harrow

A tine harrow consists of a series of rigid, pointed teeth that are affixed to a frame. These tines are generally made of metal and are evenly spaced along the width of the harrow. When the harrow is pulled by a tractor or other agricultural machinery, the tines penetrate the soil, breaking up clumps and leveling the surface.

This action helps create a finer seedbed for planting, enhances soil aeration, and facilitates water infiltration.

Spike Tooth Harrow

The spike tooth harrow features teeth that resemble long spikes, designed to stir the soil. It is also referred to as a peg tooth harrow, drag harrow, section harrow, or smoothing harrow. Its primary function is to smoothen and level the soil immediately after ploughing. The harrow works to a depth of about 5 cm when weighted. It is used in cultivating crops such as corn, cotton, and other row crops during their early stages of growth. The sections typically range from 1.2 to 1.7 meters in width and may have 25, 30, or 35 teeth. Multiple sections can be connected to a hitch bar, enabling the harrow to cover a wide swath. These sections can be either rigid or flexible. When the sections feature guard rails across the ends of the bars, they are known as close-end harrows and those without guard rails are called open-end harrows.

Animal-Drawn Spike Tooth Harrow

The animal-drawn spike tooth harrow is generally a rigid type, though it may or may not allow for adjusting the angle of the spikes during operation. This harrow is composed of teeth, a tooth bar, guard rails, clamps, braces, levers, and draft hooks. The teeth are made of hardened steel and come in square, triangular, or circular sections. The teeth are arranged on the tooth bar in such a way that no tooth is directly behind another. The teeth are securely attached to the tooth bar, and the clamps are fixed to prevent any loosening during use.

Figure 2.3: Spike Tooth Harrow - Animal Drawn

Spring Tooth Harrow

Spring tooth harrows are constructed in sections similar to spike tooth harrows. The sections typically range in width from 0.9 to 1.7 meters and can have between seven and twelve teeth. A single section can be used independently, or multiple sections can be hitched together to form a unit, depending on the available power.

The teeth of the harrow are made from wide, flat, curved, oil-tempered bars of spring steel. One end of each bar is securely attached to a bar, while the other end is pointed to ensure effective penetration into the soil. The depth of penetration is adjustable by modifying the angle of the teeth using levers. Some spring tooth harrows are equipped with a power-angling hitch for further adjustability.

Figure 2.4: Spring Tooth Harrow

Spring tooth harrows are adapted for use in rough and stony ground. They are also used extensively to loosen previously ploughed soil ahead of grain drill seeding rice or small grains. The teeth will penetrate deeper than those on spike tooth harrow, and they will give when obstructions are struck. Spring tooth harrow is also called as quack grass, and Bermuda grass eradicator, since the teeth penetrate deeply, tear out and bring the roots to the surface. Spring tooth harrows are available for animal and tractor operation

Acme Harrow

It is a unique type of animal-drawn harrow featuring a transverse horizontal frame with rigid, curved blades. It is also referred to as a blade harrow, curved knife-tooth harrow, or pulverizer, the front portion of the blades breaks up the soil and crushes the clods. This harrow is effective in achieving fine soil pulverization and is ideal for creating a soil mulch.

Figure 2.5: Acme Harrow

Patela

It is a wooden plank designed to smoothen soil and crush weeds. It is also used for breaking clods, packing, leveling ploughed soil, and removing weeds. The harrow features a wooden plank with several curved steel hooks bolted to a steel angle section fixed to the rear of the plank. The plank's cutting edge helps level and pack the soil, while the curved hooks are effective in uprooting and collecting the weeds. This version is drawn by animals.

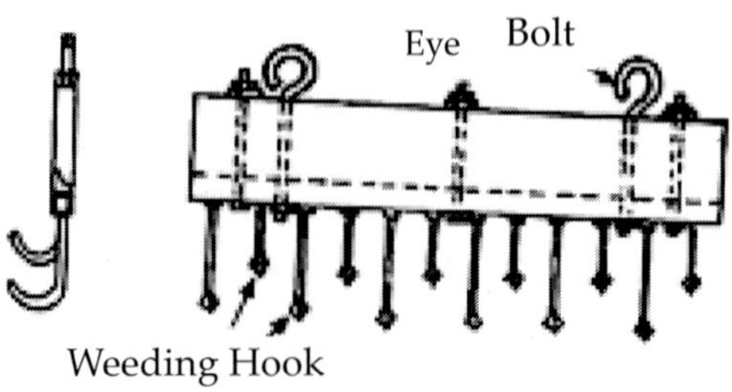

Figure 2.6: Patela

TRIANGULAR HARROW

It is a spike tooth harrow with a triangular wooden frame, into which pointed spikes are fixed. The teeth of the spikes are stationary and cannot be adjusted. This harrow is primarily used for breaking clods and smoothing the soil surface.

Drag Harrow

A drag harrow consists of multiple metal bars arranged in a grid-like pattern, all attached to a frame that can be towed by a tractor, ATV, or other suitable vehicles. As the harrow is pulled across a field, its tines break apart soil clumps, create a smooth surface, and evenly distribute leftover crop materials. Drag harrows are particularly effective for light tilling, smoothing uneven landscapes, and incorporating organic matter into the soil.

Blade Harrow

Blade harrows serve various agricultural functions, including weed and stubble removal, clod crushing, shallow soil cultivation, seed covering, and inter-cultivation harvesting. They are divided into two main types which are indigenous and improved blade harrows.

Indigenous blade harrows are traditional tools, typically locally crafted, used for basic cultivation tasks. In contrast, improved blade harrows are modern versions designed with enhanced materials and engineering, increasing their efficiency and performance. This animal-drawn implement is primarily used to prepare seedbeds in clayey soils and to create soil mulch for moisture conservation. It typically consists of one or more blades attached to a beam, working at shallow depths with minimal or no soil inversion.

A power harrow tills the soil while maintaining the original field profile. It breaks up both the upper and lower layers of soil without inverting them, resulting in a well-prepared seed bed and effective soil mulch. The harrow consists of two horizontal crossbars equipped with rigid pegs, which moves back and forth, powered by the tractor's PTO. The pegs are spaced 200 mm apart and are staggered across each crossbar. As the two bars move in opposite directions, the implement remains dynamically balanced. The oscillating pegs break up clods and finely pulverize the soil. The operational width is 2000 mm, and the field capacity is approximately 1.5 hectares/day.

ROTARY TILLER

The rotary tiller, also known as a rotary cultivator, is regarded as one of the most important agricultural tools due to its ability to provide a fine level of soil pulverization. It is directly mounted to the tractor for operation. The key advantages of using a rotary tiller are: a) effective soil pulverization that promotes healthy plant growth, b) efficient cutting and mixing of stubble and roots with the soil, and c) field leveling.

The main components include a rotor equipped with 36 to 48 L-shaped steel blades, a gearbox, power shaft, sprocket-chain drive, universal joint, leveling board, shield, depth control system, and a three-point hitch. Power is transferred from the tractor's engine to the rotary tiller through the PTO (Power Take-Off). A leveling board attached to the rear helps in levelling the tilled soil. Two adjustable brackets on either side of the unit control the tilling depth. The rotor operates at 180-200 rpm.

Common Types of Blades Used in Rotary Tillers

- **'L' Type Blade:** Performs well in trashy conditions, effectively cutting weeds but with less soil pulverization.
- **Twisted Blade:** Ideal for deep tillage in relatively clean ground, but it can clog and require frequent cleaning due to trash buildup on the tines and shafts.
- **Straight Blade:** Typically used in mulchers designed for secondary tillage.

Function & Operation of Rotary Tillers

- **Function:** These machines use a rotating set of blades or tines to till the soil, breaking it up into fine particles. They are often used to prepare the seedbed by thoroughly mixing soil and organic matter.
- **Operation:** As the tiller rotates, its blades chop and mix soil, which helps to improve soil structure, break up compacted soil, and mix in crop residues or compost.
- **Best For:** Loosening compacted soils, mixing organic matter, and creating a fine, even seedbed for small-seeded crops.

LAND ROLLERS OR PULVERISERS

Land rollers or pulverisers are tools used for further preparation of seed bed.

- **Clod Crusher:** It is used to finalize the preparation of the seedbed by finely pulverizing and compacting the loose soil, ensuring there are no large air spaces or pockets. It presses the topsoil down against the subsoil, creating a uniform seedbed that helps conserve moisture and delivers it to the plant roots as required.

Figure 2.7: Single Gang Land Roller (Clod Crusher)

- **V- Shaped Roller Pulverizer:** It is made up of several wheel sections mounted on a shaft. The surface of the roller features a corrugated pattern, which is why it is also referred to as a corrugated roller. It performs multiple tasks including rolling, pulverizing, packing, leveling, cultivating, and mulching the soil.

Levelers

Land leveling is a permanent improvement done on agricultural lands. It is carried out to smoothen the contours of land to achieve certain desirable objectives for efficient agricultural production. Land leveling helps for (i) efficient application of irrigation water, (ii) improved surface drainage, (iii) minimizing soil erosion (iv) conservation of rain water in dry lands and The animal drawn leveler consists of a wooden leveling board with a handle . In the front portion of the board two hooks are provided for connecting it to the yoke of the animals.

SOIL SCOOP

Soil scoops are used for tasks such as digging ditches, clearing drains, and performing cut-and-fill operations in land leveling. The implement consists of a trough, a blade, a hitching loop, and a handle. Soil scoops are available in both animal-drawn and tractor-drawn versions.

- **Blade:** The blade is crafted from light carbon steel, with a carbon content ranging from 0.5% to 0.6%. The cutting blade is set at an angle between 12° and 15°, and it is either riveted or bolted to the soil trough.
- **Soil Trough:** Made of mild steel sheet, the trough is shaped into a trough form to carry soil.
- **Hitching Loop:** The loop is made from from mild steel, with the ends attached to the side of the soil trough. It includes iron rings or pins to connect the scoop to either animals or a tractor.
- **Handle:** In the animal-drawn version, the scoop is fitted with two handles made of timber or mild steel plates. These handles are attached to the hitching loop and are used to control the movement of the implement.

It is a tractor-mounted tool operated through the tractor's hydraulics and three-point linkage. The implement includes a hitch system, a sharp-edged, replaceable cutting blade, and a curved plate with side wings that together form a bucket. During use, the blade penetrates the soil, collecting excess soil in the bucket, which is then deposited into field depressions. The leveler's angle and pitch can be adjusted, and it can also be tilted to the left or right, or reversed for backfilling.

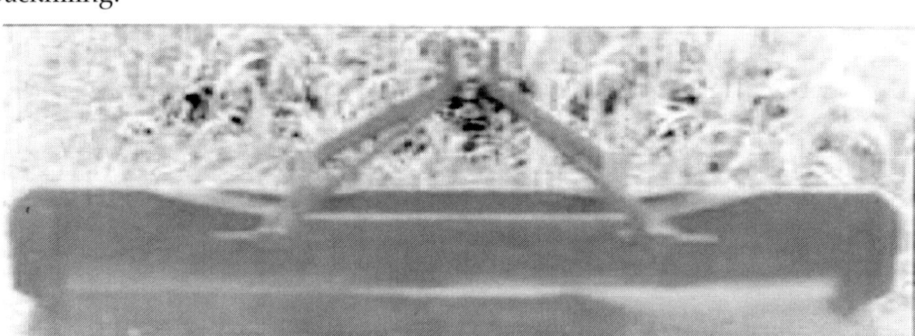

Figure 2.8: Tractor Drawn Leveler

- **Leveling Board- Bullock Drawn:** It is a wooden board of length 2.0m, width 0.4m and thickness 0.3m provided with side wings, hitching braces and handle. The board is hitched to animals and pulled. The blade scrapes soil from elevated places, the soil is moved to places where there are depressions and released. It can be used in leveling of paddy field after puddling

CULTIVATORS

Cultivators are mainly used to loosen and aerate the soil, as well as control weeds. They help in seedbed preparation by improving soil structure and ensuring nutrient availability.

Types of Cultivator

- **Spring-tooth Cultivator:** Features spring-loaded teeth that allow for deep soil penetration while remaining flexible to avoid damage from rocks or other obstacles.
- **Rigid Tine Cultivator:** Uses fixed, non-flexible tines to break up the soil and eradicate weeds, typically suited for lighter, loamy soils.
- **Rolling Cultivator:** Equipped with rolling wheels that stir the soil and incorporate organic material, ideal for preparing a fine seedbed.

Benefits of Secondary Tillage

- **Enhanced Soil Structure:** Secondary tillage helps in breaking down large soil clumps, improving aeration, and refining the seedbed structure.
- **Weed Control:** Many secondary tillage tools uproot or disturb weeds before planting, aiding in their control.
- **Organic Matter Incorporation:** It mixes organic matter, such as crop residues, cover crops, or compost, into the soil.
- **Better Seedbed Preparation:** It creates a uniform and fine seedbed that supports seed germination and root development.
- **Improved Water Infiltration:** Breaking up compacted soil improves water infiltration and reduces the chances of waterlogging.

Limitations of Secondary Tillage

- **Time and Cost:** Secondary tillage requires additional time and equipment, which can increase costs.
- **Soil Compaction:** Excessive use of certain secondary tillage tools, especially heavy ones, may lead to soil compaction if not managed properly.
- **Erosion:** Frequent tillage in areas with inadequate erosion control can disturb the soil structure, increasing the risk of erosion.

Final Thoughts

Secondary tillage is essential in crop cultivation, refining soil conditions after primary tillage. It aids in seedbed preparation, weed control, the incorporation of organic matter, and overall soil conditioning, ensuring optimal growing conditions for crops.

3

Weeding Equipment

An agricultural weeder is a tool designed to remove weeds from fields or gardens. It can be used manually or attached to machinery like tillers and power weeders. Weeding is necessary to eliminate unwanted plants that compete with crops, ensuring better growth and profitability. Effective weed management is crucial in crop production, as weeds can reduce yields by stealing nutrients and attracting harmful pests. In recent years, research has focused on non-chemical methods and environmentally friendly alternatives to herbicides for weed control. Mechanical weeding is often the most cost-effective solution. Tools like rotary weeders, mechanical hand weeders, and wetland weeders are generally effective, with wetland versions gaining popularity due to their ability to function in shallow water. These tools uproot weeds and bury them in the mud through their pull-and-push action, making them both safe and eco-friendly. Weeding should be done at least three times during a growing season, with early weeding being the most effective for completely eliminating young weeds.

FUNCTIONS OF WEEDING EQUIPMENT

Secondary tillage equipment is often used for weed management. Weeders help control weeds between planted crops or before sowing. Some weeding machines use mechanical methods to remove unwanted plants without the need for chemicals.

Usage Instructions

- Early weeding is essential: Perform the weeding within 20 days after sowing.
- Delayed weeding can lead to competition between paddy and weeds, harming the crop.
- Move the weeding machine back and forth between rows.
- Conduct a second round of weeding after 15 days, and if possible, a third.

- While the third weeding is optional, each weeding helps aerate the soil and provides significant benefits.
- A fourth weeding before flowering can further enhance yield, even if the field is mostly free of weeds.

TYPES OF AGRICULTURAL WEEDERS

Agricultural weeders are tools or machines designed to eliminate or control weeds in fields, preventing competition with crops for nutrients, water, and light. Effective weed control is important for maximizing crop yields. Agricultural weeders range from manual tools to large-scale machinery. Some common types include:

- **Handheld Weeders:** Small, manual tools used to remove weeds from gardens or fields. These are typically made of metal or plastic and come in various shapes and sizes.
- **Hoes:** A traditional tool used to cut, chop, or dig up weeds at or below the soil surface. Hoes are effective for both shallow and deep-rooted weeds, featuring a long handle with a flat, blade-like end.
- **Hand Hoe:** The most commonly used manual weeding tool in farming. It has an iron blade and a wooden handle, typically 30-40 cm in length. The operator uses it to cut the soil to a shallow depth of 2-3 cm, removing weeds while mixing the soil. The hand hoe covers 5-7 cents of land per day.

Figure 3.1: Hand Hoe

- **Hoe-Cum-Rake:** The Hoe-Cum-Rake is a versatile hand tool designed for multiple purposes. It features a flat blade on one side and prongs on the other, both made from a single piece of metal with an eye in the center or welded to an eye. A wooden handle is attached for ease of use. The flat blade is used for digging, raking, weeding, and collecting weeds and debris, while the pronged end is used for raking. This tool is usually used in secondary tillage operations, especially for preparing nursery beds, and is typically employed for lighter tasks. The flat side of the tool works through impact action, while the rake end is used for gathering.

Figure 3.2: Hoe-Cum-Roke

- **Cultivators:** Cultivators are machines designed to loosen the soil and eliminate weeds from a field. They can be either tractor-mounted or used as standalone equipment, depending on the needs of the farmer.
- **Rotary Hoes:** Rotary hoes are machines equipped with spinning blades that cut weeds off at the base. They are versatile and can be used in both garden settings and larger fields.

Figure 3.3: Rotary Hoes

Long Handle Weeders

Hand hoes require more physical strength because their short handles place additional stress on the operator. As a result, the operator must perform the weeding task while bending, which can be physically demanding.

Figure 3.4: Long Handle Weeders

To prevent the need for bending during use, hoes now come with long handles, which are referred to as long-handled weeders. The most common types of long-handled weeders are:

- Star-type weeders
- Peg-type weeders

These weeders are primarily used in dry soil conditions, which is why they are called as dryland weeders.

Star-Type Weeder

Star-type weeders are ideal for dryland weeding and are usually used in garden fields, particularly when soil moisture content is low, around 10-15%. However, they are most effective in line-sown crops and not in broadcasted fields.

This weeder features a blade for cutting weeds and a fulcrum wheel that allows for a push-pull motion. The long handle reduces strain on the operator, making the task easier. The fulcrum wheel's radial arms are designed with star-like projections, giving its name as a star-type weeder. The starwheel design is most effective in loamy soils, and the blade typically has an operating width of 120mm, with a coverage of 0.05 ha/day. These weeders are also referred to as dryland weeders.

Figure 3.5: Star-Type Weeder

Peg-Type Weeder

The peg-type weeder is designed for dryland soils and is most effective in conditions where the soil moisture is low, around 10-15%. It can also be used in garden fields under similar conditions. However, it is best suited for line-sown crops and is not effective in broadcasted fields. The weeder consists of a blade for cutting weeds, a fulcrum wheel for push-pull movement, and a long handle to ease operation and reduce strain on the user. The peg-type weeder gets its name from the pegs welded onto its wheel, which is designed for use in clayey soils. The operating blade width is 120mm, and its coverage is 0.05 ha/day. These weeders are also known as dryland weeders.

Wheel Hoe

The wheel hoe is a widely used tool for weeding and intercultural work in row crops. It is a long-handled tool operated by a push-pull motion. The wheel hoe consists of a wheel, tool frame, a set of replaceable tools, and a handle. It can be equipped with various soil-working tools, such as straight blades, V-blades, sweeps, and shovels, making it versatile for different tasks.

Figure 3.6: Wheel Hoe

Wheel hoe can be used for different works like weeding, soil mulching, stirring, and so on. The long handle reduces drudgery for the operator. Wheel reduces the energy requirement for pushing. The tools of the soil working components are made from medium carbon steel. Its coverage is 0.05 ha/day.

Cono Weeder

The Cono weeder is primarily used in wetlands to uproot and bury weeds in line-planted rice fields. It helps disturb the topsoil, enhancing aeration and creating a more favorable growing environment for the crop.

Figure 3.7: Cono Weeder

This weeder features a long handle, two truncated conical rollers, and a float. The rollers are positioned at the bottom of the handle in opposite directions, one behind the other, and are equipped with serrated blades along their edges. During operation, the weeder is moved between two rows of standing crops, where the rollers uproot and bury the weeds. The action of the weeder also stimulates root growth. For optimal use, the soil should have adequate moisture and be slightly firm. The weeder covers an area of 0.05 ha/day.

USES OF AGRICULTURAL WEEDERS

Agricultural weeders have several important uses across different sectors, including:

- **Crop Production:** One of the primary applications of agricultural weeders is in crop production. Weeds can significantly hinder crop growth and reduce yields, making it essential to remove them to ensure optimal production.
- **Garden Maintenance:** Agricultural weeders are commonly used in garden settings, helping to clear weeds from flower beds, vegetable patches, and other areas.
- **Organic Farming:** In organic farming, where the use of synthetic herbicides is restricted, agricultural weeders are immensely valuable. Organic farmers rely on mechanical methods, such as weeders, to control weed populations and preserve the integrity of their organic certification.
- **Environmental Conservation:** Weeds can negatively impact the environment, so using agricultural weeders to remove them helps preserve the natural ecosystem and maintain biodiversity.

Advantages of Using Agricultural Weeders

Agricultural weeders offer several advantages, including:

- **Reduced Need for Herbicides:** Agricultural weeders help reduce the reliance on herbicides, which can be harmful to the environment and costly to apply.
- **Improved Crop Yields:** By eliminating weeds from fields or gardens, agricultural weeders enhance crop yields and boost overall production.
- **Reduced Labor Costs:** Agricultural weeders help cut down on labor costs by automating the process of weed removal.
- **Increased Efficiency:** These tools improve the efficiency of weed removal, ensuring that weeds are eliminated quickly and effectively.

BASKET WEEDERS

Basket weeders are primarily used for between-row weed control, but they do not address within-row weed management. They are particularly effective for crops such as head lettuce, where soil heaping in the rows must be avoided. These weeders are often used alongside finger weeders for comprehensive weed control.

Basket weeders consist of two sets of rotating wire cages. The front cages are ground-driven, while the rear cages are powered by a chain connected to the front cages, turning twice as fast in the opposite direction. This design ensures that even small weeds experience significant damage, even if they aren't fully uprooted. The soil flow is kept parallel to the crop row, allowing the tool to operate within approximately 2.5 inches of the crop, making it ideal for young crops. Manufacturers also design specially shaped baskets for cultivating bed sides. However, since the baskets must be sized to fit the inter-row dimensions, separate machines are required for different row spacings. Basket weeders are not suitable for stony ground, as rocks can damage the baskets and get caught between the wires.

Figure 3.8: Basket Weeder.

Basket weeders are metal cages designed to roll over the soil surface, gently scuffing it without displacing soil into the crop rows. This feature makes them ideal for newly emerged crops or crops like lettuce, which need to remain soil-free and cannot tolerate hilling. Custom-built for two to eight-row beds, Buddingh basket weeders come with angled baskets for working the sides of raised beds. The basket widths vary from 3 to 14 inches, depending on the row spacing. For wider rows or varying row widths as crops grow, overlapping baskets that "telescope" or expand and contract are available for easy adjustment.

The front baskets rotate at ground speed, while a chain drives the rear baskets slightly faster. This setup helps kick up soil and dislodge weeds that survive the first set of baskets. Typically operated at speeds between 4 to 8 mph, the tool works best with straight rows and an experienced operator to prevent crop damage. At higher speeds, both baskets provide hoeing action. The tool is usually belly-mounted to facilitate close cultivation. While it can handle small stones, it performs best in fine soils free of crop residues and is most effective on small weeds. It can also manage a dense stand of inch-high weeds, provided the soil is friable. A 3-row unit on a 4-foot frame costs around Rs.1.2 Lakhs though prices may vary depending on mounting hardware, and it's recommended to order in advance.

BRUSH HOES

Brush hoes are designed for close cultivation in narrow rows. These units are slightly expensive, starting at Rs. 5 Lakhs per row. The equipment includes shields that protect plants from the bristle wheels, which rotate independently between the rows. These wheels "sweep" small weeds out of the soil while also creating a "dust mulch" that helps suppress further weed germination. The operator sits behind the rotating wheels, steering the unit to ensure close cultivation.

The standard brush hoe is equipped with PTO-driven polypropylene brushes that work parallel to the crop row. These cultivators uproot smaller weeds and shear off larger ones. The soil is moved primarily along the row, and with narrow tunnel shields (ranging from 2.5 to 8 inches wide), it is possible to cultivate very close to small crop plants. However, as the crops grow larger and become harder to move through the shields, it becomes essential to switch to a different type of cultivator.

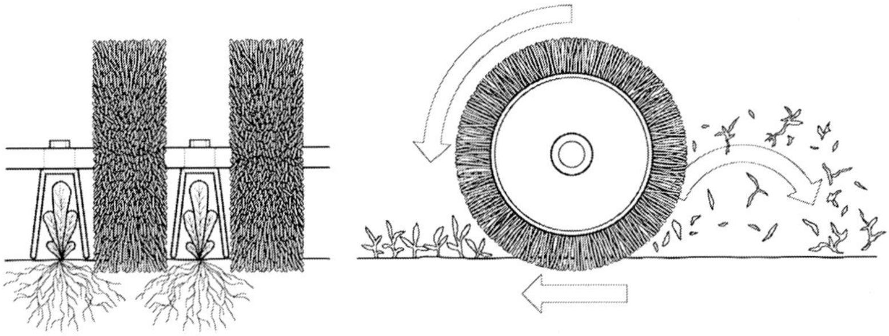

Figure 3.9: Brush Hoe. (A) Rear View; (B) Side View

The brush hoe is designed to resist clogging with large weeds and debris, making it effective in wet soils. One of its key benefits, similar to the basket weeder, is that it leaves a loose, uniform soil surface that helps slow weed germination. However, because the brushes are mounted on a common axle, the implement requires a flat seedbed to maintain consistent operating depth. As tractor speed increases, the working depth decreases, making high-speed operation impractical. Adjusting the row spacing requires significant disassembly of the tool.

Since brush hoes work very close to the crop row, they are equipped with a driver's seat and steering mechanism, making them a labor-intensive implement. Additionally, when operated on dry soil, they create a considerable amount of dust, which can make steering unpleasant for the operator.

UNDERSTANDING TORSION WEEDERS AND SPRING HOES

Torsion weeders and spring hoes are excellent for removing both small and larger weeds. The torsion weeder can be directly attached to any hoeing mechanism without the need for adjustments. It is designed to be suitable for a wide range of crops. Due to its low mounting point on the machine, the weeder does not damage the soil. For more intensive tillage, two additional teeth can be attached, bringing the total to four teeth per ridge for a deeper result. Torsion weeder is especially useful for tough soils or when a high level of weeding is required. Offers two types of 7mm teeth: one with a horizontal curl and an optional tougher version with a vertical curl.

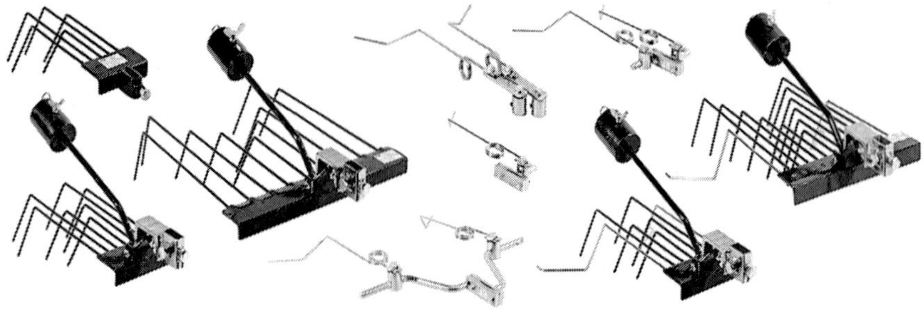

Figure 3.10: Torsion Weeders and Spring Hoes

Functions of Torsion Weeders & Spring Hoes

The teeth of the torsion weeder can be adjusted to the desired width based on the crop's growth phase. In the early stages, the weeder breaks open the ground while simultaneously destroying germinating weeds. As the crop matures, the teeth can be positioned closer together. The driver's speed and the ground resistance cause the teeth to bend in such a way that crop damage is prevented. At the same time, any germinating weeds are uprooted. The teeth create slight ridging, which helps cover the weeds within the ridge, further enhancing weed control.

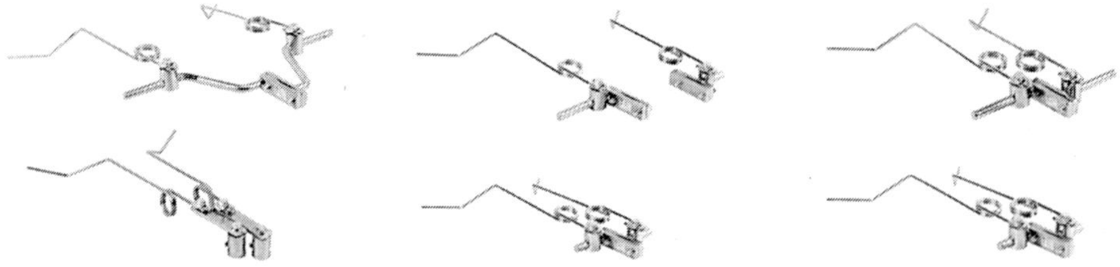

Figure 3.11: Torsion Weeder V2

The new torsion weeder consists of two clamps for attaching teeth, which are easily set up and fitted to any hoeing machine. The teeth can be adjusted using a single bolt, allowing for changes in length, angle, and position relative to the crops. This offers two types of teeth: one with a horizontal curl, which applies more pressure to the ground, and one with a vertical curl, which is directed more toward the crops.

Easily Adjustable

With clamps, the torsion weeder can be quickly adjusted. The width, angle, and depth (pressure) of the teeth can all be modified using just one bolt. This feature significantly reduces time compared to existing systems.

Additional Teeths

The torsion weeder mechanism allows for an additional set of teeth to be mounted, enabling more intensive tillage. This setup adds two horizontal and two vertical teeth per element.

Weed Reduction

Studies have shown that a weed reduction of 69% is achievable with torsion weeders alone. This can be increased to 90% when torsion weeders are used in combination with finger weeders.

VERTICAL AXIS TINE AND BRUSH WEEDERS

Tine weeders are excellent for cultivating a stale seedbed and for use when crops are slightly larger, typically at the 3-4 leaf stage or beyond. In a stale seedbed, they lightly scratch the soil surface, uprooting small, newly germinated weeds while providing good soil coverage. However, when crops are still small, especially direct-seeded crops, tine weeders can be less precise, potentially uprooting the crops themselves. For instance, a demonstration in a beet field, where the beets were around 2.5 inches tall, showed that while some survived, others were uprooted. While this method can thin crops like beets, it may not be the most accurate. However, once crops have developed stronger root systems, this tool is very effective, particularly for crops like beets or carrots with deeper taproots.

Figure 3.12: Vertical Axis Tine and Brush Weeders

These machines are designed to work very close to, or sometimes within, the crop row. Rear-mounted, they feature a seat for the operator, though the tine-type models can be locked for inter-row cultivation without the need for an operator. The operator controls the position of rotating wire baskets or polypropylene brushes, which stir the soil next to the crop. Handles allow the operator to adjust the tools' position accurately. In crops with sufficient spacing, the operator can position the tools fully within the crop row, effectively removing most in-row weeds. These are precision tools that require skilled operators for optimal performance while reducing crop damage. The speed of cultivation largely depends on the operator's expertise.

The tine-type machines consists of inverted baskets with nearly vertical spring tines that aggressively stir the soil, uprooting and breaking moderate-sized weeds. These machines are particularly beneficial in strawberry production, where they control weeds and also brush runners into the crop row with minimal damage. Since the operator independently controls the position of each rotating basket, the standard machine is designed to cultivate a single row. However, similar machines equipped with a machine vision system can cultivate multiple rows simultaneously.

The brush-type machines have inverted cones with bristles that gently brush the soil surface, causing minimal disturbance to the soil profile. While they can uproot very small weeds, their primary action is shearing off weeds at the soil surface. Because the bristles are flexible, they can brush against the stems of certain crops with little to no damage, allowing the brushes to work effectively within the crop row to control small weeds.

SPINNERS

Normally, these are used in a staggered pair with the two spinners scratching across the row in opposite directions. These ground-driven in-row weeders feature a basket-like arrangement of spring steel wires that scratch laterally across the crop row. Spinners are typically used in pairs, with each tool working in opposite directions across the row to increase the area worked. The depth is generally set so that the deepest penetration is slightly above the planting depth, directly in the row. Alternatively, the tool can be adjusted to penetrate deepest next to the row, scratching across the row with an ascending or descending motion. In any case, deeper planting allows for more aggressive weeding. Cultivating late on a sunny day, when crop stems are less stiff, helps in reducing mortality in crops like corn. Spinners are most effective against weeds in the white thread to cotyledon stages and are safest when the crop is still small. As a result, their usefulness overlaps significantly with that of rotary hoes and tine weeders, which are easier to use but less versatile.

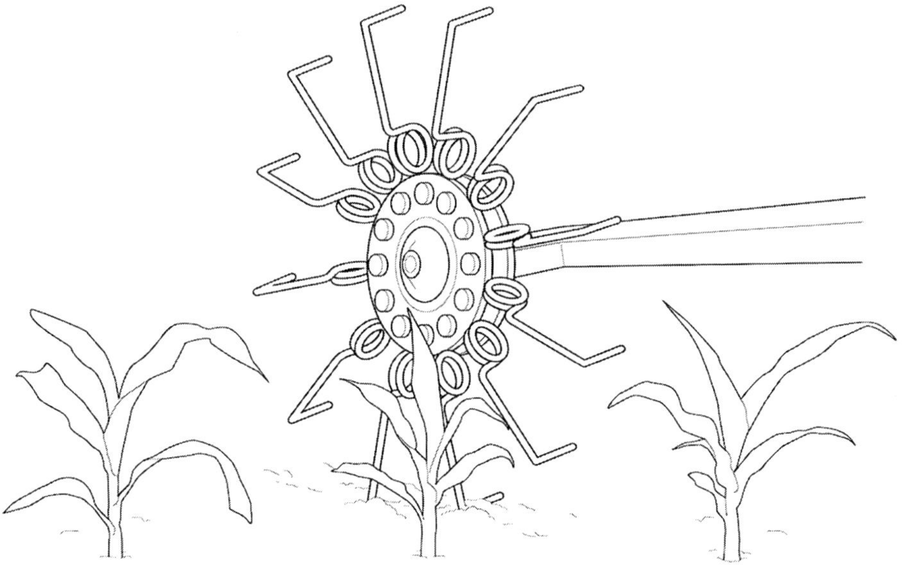

Figure 3.13: Spinner

A direct strike by the tip of a wire can destroy or cut off a crop plant, so spinners tend to thin closely planted crops like soybeans or peas. However, with proper tool adjustment, mortality typically remains under a few percent in crops spaced more than 6 inches apart in the row. Spinners should not be used in transplanted crops until the plugs are well rooted. When set deep, spinners can be used intentionally to thin an overly dense crop stand. In dense stands, the risk of damaging plants is higher, but thinning the stand with spinners can lead to a more uniform crop spacing.

OVERVIEW OF RUBBER FINGER WEEDERS AND RUBBER STAR WHEELS

Rubber Finger Weeders, also known as Buddingh 'C' cultivators, are designed for usage around the stems of sturdy crop plants that can tolerate some contact. These tools use rubber-coated metal fingers for in-row weeding, which are connected to a deeper set of metal fingers that drive the unit at ground speed. These weeders operate at low speeds, typically just a few miles per hour, as they work closely to the plants. They require belly-mounting and are best suited for use with a G-type tractor. In wet, clayey soils, the fingers can become clogged and require frequent cleaning.

Rubber Finger Weeders

Rubber finger weeders consist of two pairs of ground-driven wheels, each equipped with rubber-tipped fingers that stir the surface soil in the row while bending around well-rooted crop plants. Behind these, wire baskets aggressively stir the area adjacent to the row. Typically, the implement is belly-mounted. Single-row machines can work with row spacings as narrow as 20 inches, but cultivating multiple rows requires a minimum row spacing of 34 inches. This effectively limits the machine's use to high-value crops that justify cultivating one row at a time or to crops with wide row spacing. While the rubber fingers have some flex, they are firm enough to damage most young crops. Unlike other implements that attack weeds within the crop row, finger weeders can be used on large, upright crop plants, with the upper height limit set by the clearance under the tractor, making them useful in nurseries.

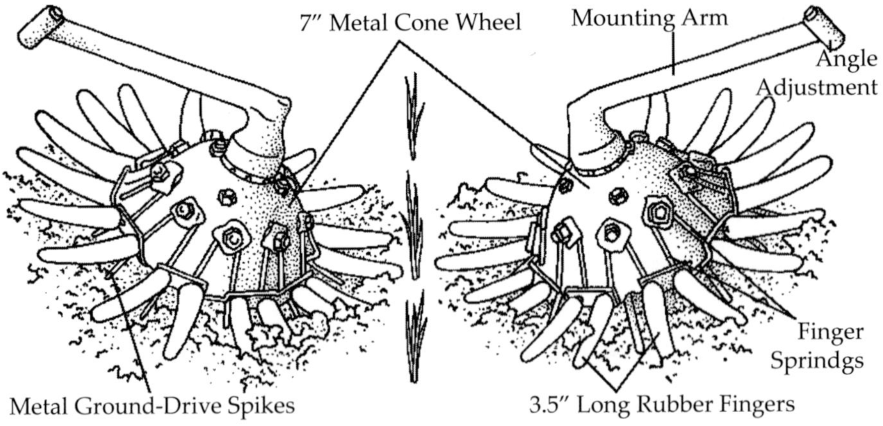

Figure 3.14: Rubber Finger Weeder

Adjustment of the Finger Weeders

Finger weeders should be set with a distance of approximately 2 cm. In practice, achieving this adjustment is only possible when the machine is equipped with a camera steering system (ROW-GUARD). On machines without camera steering, whether front or rear-mounted, achieving a precise 2 cm setting is difficult to see and consequently hard to achieve. As the crop grows larger, the distance should be set wider as the crop or its roots become thicker.

The finger stars should slightly bend upwards with light pressure applied. During operation, the spring on the holder should be under slight pressure to allow the fingers to work effectively and adapt to uneven ground. This can be adjusted by changing the height of the shaft of the finger weeders. The spring also compensates for any frame movements, which may occur due to support or feeler wheels on the cultivator.

The steel drive fingers, which penetrate the soil, must work at the correct depth—not too deep or shallow. The depth depends on soil type, soil conditions, and the crop stage. The height of the finger weeder should be adjusted to ensure it works shallow, but most importantly, not too deeply into the soil.

It's important that each crop row has a separate spring-mounted holder for the stars, which are mounted in pairs. If the stars were mounted individually on separate holders, different lateral pressures would be applied to the plant row if the heights differed. Additionally, the finger weeder must work the row parallel and without offset to ensure precise operation without damaging the crop. Finger weeders that are not placed opposite each other can harm the crop, operate at different heights, and move or push the crop, reducing the weeding effect, as the fingers don't work on both sides of the row simultaneously.

Figure 3.15: Rubber Star Wheels.

Rubber Star Wheels

Rubber star wheels are conceptually similar to rubber finger weeders but are specifically designed to mount on a row-crop cultivator to provide in-row and near-row weeding that sweeps and shovels alone cannot achieve. These rubber stars come in various stiffness grades, with more flexible versions being suitable for use in young and delicate crops. The looser the soil in the row, the better the tools work when more flexibility is present in the rubber. As with torsion weeders and spring hoes, determining an appropriate mounting location on a cultivator gang can be difficult. Like all in-row weeding tools, precise guidance is key to avoid crop damage and ensure effective weeding in the row.

Pneumatic Weeders

Pneumatic weeders consist of a PTO-driven air compressor connected to a pair of small air-blast nozzles positioned in each row. The nozzles point toward the row but slightly backward, traveling just beneath the soil surface. The air blast is effective at dislodging shallowly rooted weed seedlings without harming more deeply rooted crop plants. For optimal performance, the soil must be loose and friable.

ROTARY HOES

Rotary hoes are a more aggressive form of blind cultivation compared to tine weeders. These hoes consist of thin spider wheels, typically 16 or 18 inches in diameter, spaced 3½ inches apart across the unit. The spider wheels turn independently and support the weight of the unit, with optional gauge wheels available. Rotary hoes are priced around Rs. 1.5 lakahs for a 6-foot width, Rs. 2.2 Lakhs for a 12-foot width, and can be found in sizes up to 24 feet. Initially designed for blind cultivation of grain crops, rotary hoes are effective for controlling small weeds in newly emerged corn or beans. They excel at breaking up crusted surfaces of soil and perform phenomenal in heavy soils, but are not suitable for lighter soils, as they can work too deeply and become too heavy. Rocks can clog the wheels, preventing them from turning correctly and potentially damaging the crop. Plastic mulch pieces may also collect on the wheels and need to be removed. Dull spider tips can reduce the effectiveness of rotary hoeing.

A rotary hoe consists of one or two rows of wheels, each with 16 spoon-like projections. The wheels are attached to the toolbar via spring-loaded arms that allow movement over obstacles. The ground-driven wheels typically penetrate the soil to a depth of 0.8 to 1.5 inches and flick up soil and small weeds as they rotate. To be effective, the machine should be operated at high speed, enabling rapid treatment of large areas. The shallow penetration allows the wheels to make contact with young crop plants with minimal damage.

Most rotary hoes under 20 feet in width do not come with depth gauge wheels. If rotary hoe lacks these wheels, it is advisable to acquire them. While excessive soil penetration is not a frequent problem, operating a rotary hoe deeper than the planting depth can cause significant damage to the crop. Using the three-point hitch to carry the implement is not a suitable alternative to gauge wheels, as it causes the tractor to bounce at the high speeds required for effective operation, leading to inconsistent penetration.

Rotary hoes are most effective at killing weeds in the white thread stage. By one or two days after emergence, most weeds are sufficiently rooted to survive. Therefore, timing is crucial for success. Typically, the implement is used five days after planting, with additional passes at five- to seven-day intervals. Generally, two or three passes are all that can be made before the crop becomes large enough to sustain damage. One study showed that three timely rotary hoeings reduced weed dry weight by an average of 72% in drilled soybeans, while two untimely passes only reduced dry weight by 38%. However, most of the weed control comes from the first pass if the timing is correct.

Figure 3.16: Rotary Hoe

Wet soil conditions can decrease the effectiveness of rotary hoeing. When the soil is too wet, the spoons merely punch holes in the ground instead of tossing the soil and seedlings into the air. Additionally, rainfall or irrigation shortly after rotary hoeing reduces the percentage of weeds that dry out and die.

Unlike many other cultivating implements, rotary hoes have a limited lifespan. The spoon tips wear down over time, requiring replacement of either the tips or the entire wheels. Worn tips are not effective at throwing soil, reducing the implement's overall effectiveness.

Merits of Using a Rotary Hoeing

The rotary hoe is effective for controlling weeds that appear shortly after planting, especially when there hasn't been enough rainfall to activate preemergence herbicides. It creates minimal disturbance to crop residue, which helps improve soil infiltration and decreases the risk of erosion. Additionally, the cost is relatively low, with fuel consumption for rotary hoeing at only 0.20 gallons of diesel per acre.

TINE WEEDERS (WEEDING HARROWS)

Tine weeding is a highly efficient method of mechanical weed control. When used properly, a tine weeder can eliminate up to 70% of weeds. Achieving 100% weed control is unrealistic, however. The main objective of using a tine weeder is to target weeds before they begin to grow, giving the crop a head start. While some weeds may be beneficial due to their higher density and the variety of wild herbs they provide, which contribute to soil stability.

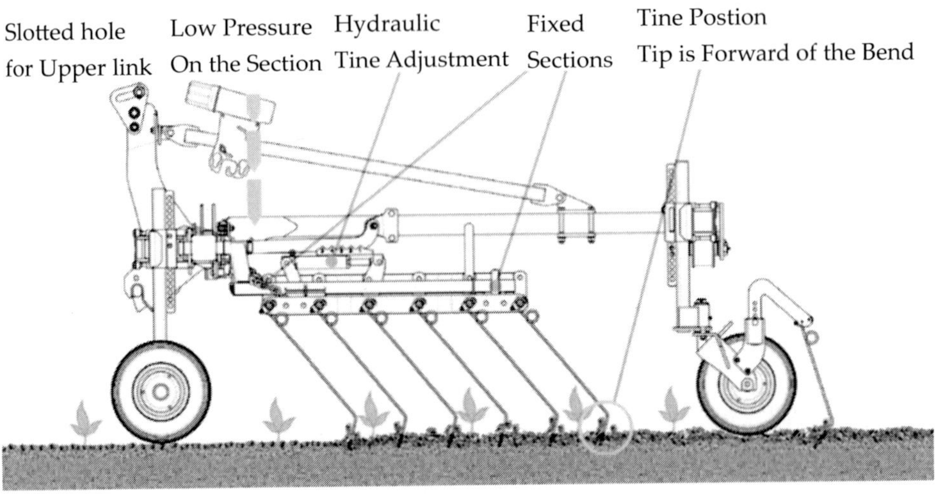

Figure 3.17: A Tine Weeder

These implements come in various designs but generally consist of a frame with numerous small-diameter tines pointing downward. They are essential for weed management on many organic grain and vegetable farms. While some farmers still use traditional chain harrows and spike or peg tooth harrows for weed control, these older methods are increasingly being replaced by tine weeders. The tines of these weeders are typically flexible spring steel wires, ranging from 0.19–0.38 inches in diameter. The appeal of tine weeders lies in their ability to adjust the tine angle and down-pressure which in turn provides versatility in aggressiveness, as well as their ability to flex around well-rooted crop plants. The springy nature of the tines also causes them to vibrate side to side, helping dislodge weeds from the soil. Unlike traditional harrows with rigid tines, tine weeders do not fling rocks onto crops. They are also sometimes referred to as "finger weeders," though they should not be confused with rubber finger weeders.

Tine weeder designs vary. Some machines have tine carriers rigidly mounted to the frame, transferring weight from the tractor to push tines into the soil for more aggressive action, but they require precise adjustment to prevent crop damage. Others suspend the tine some from chains, allowing them to float on the soil surface and reduce crop damage. Certain models combine both features, providing floating tines and the option to transfer weight when necessary. The flexibility of the tines increases with their length and decreases with their diameter, also influenced by the type of steel and coil design. A highly flexible tine will deflect easily around crops but may leave many weeds undisturbed. Tines can be straight or bent between 45 and 85 degrees. A 45-degree tine works well in most conditions, while an 85-degree tine is more aggressive being ideal for weeds in crops like green beans and soybeans but can damage crops with diffuse root systems, such as corn. Straight tines tend to be less aggressive compared to bent ones of the same flexibility, but a stiff straight tine can bury weeds in damp soil, unlike a more flexible bent tine.

For crops like corn and soybeans, tine weeding is most effective from pre-emergence to about 7 inches in height. However, soybeans should not be tine weeded between ground cracking and when seed leaves are horizontal. Tine weeding works best on weeds that are just emerging or in the white thread stage, which is why it should be done before weeds become visible. If the soil is too dry for weed seed germination, tine weeding may have little effect. Also, soil disturbance from tine weeding can prompt the germination of new weeds. If the first tine weeding is done too early, new weeds might emerge when the crop is more succesptible to damage, so timing it right is crucial. While tine weeding early by a day is better than doing it too late, it involves balancing weed emergence, crop development, weather, and soil conditions.

We have had success tine weeding soybeans until the second leaves with three leaflets begin to emerge. Once soybeans have deep taproots, tine weeding can be more aggressive. This can flatten many weeds, which can then be buried by immediately applying 1 inch of soil into the crop row with a row-crop cultivator. When corn is between emergence and about two leaves, tine weeders should be set to low aggression to avoid crop damage.

Regardless of the crop or weeder, the machine must be carefully adjusted to get rid of weeds while protecting the crop. The tine carriers should be level when in the soil, and the tines should not penetrate deeper than the seed depth, typically running at least 0.5 inch above the seeding depth (or shallower in rough or crusted soil conditions). In small grains, tine weeders are usually used before emergence, at emergence, and again when the crop has two to four leaves. Many experienced operators recommend waiting until oats have two to three leaves, and barley, wheat, or spelt have four leaves before post-emergence tine weeding. However, research shows that early weeding can be done without causing stand loss, although the safety of early post-emergence weeding depends on soil type, moisture, and the care taken in adjusting and operating the weeder. While waiting for crops to become well-established is the safest option, it also allows weeds to become larger and harder to control. Tines bent at 45–60 degrees are generally ideal for small grains. An 85-degree tine tends to uproot the grains, and a straight tine works well pre-emergence but tends to bury more plants during early post-emergence weeding than a bent tine. Some growers also use tine weeders to remove sprawling and vining weeds like chickweed and catchweed bedstraw from small grains just before the grain stems become longer.

Several studies have shown that weed control in small grains is closely linked to the percentage of grain plants covered by soil. To mitigate this, increasing planting density can help compensate for plant loss during tine weeding. Close spacing also helps deflect the tines away from the row, reducing crop damage and increasing weed control. Despite the tradeoff between crop damage and weed control, studies have found no consistent increase in yield from tine weeding in small grains and legumes planted in narrow rows, except when weed pressure is high. This does not mean tine weeding is ineffective, as it helps reduce weed seed production, which can reduce future weed problems.

To prevent damage to your tine weeder, always lower it when moving forward and avoid turning with the implement in the ground, as this can bend the tines. Tines wear faster on the sides than on the front or back, so sharpening may be needed. A sharpened tine can cut the crop and lose effectiveness at burying weeds. The flattening of the tines will also hinder their side-to-side movement, which is important for uprooting small weeds. When tines become overly sharpened, they can be clipped, but this will alter the tine length and performance. A shorter tine is stiffer and more aggressive, so adjustments should be made carefully until the new weeder's performance is understood.

Usage of Tine Weeder

During Transition Period

During your first few years of transitioning, weeds won't be as prevalent due to prior use of pesticides.

Weather

You should try to tine weed on a sunny and windy day, as the weeds will dry out quickly and die.

Tines

Tines are angled so the tines have light pressure on the soil and tine position is set so the "tip is forward of the bend". This means the bottom of the tine is slightly pointing forward into the soil in the direction you are going forward.

Speed

The aggressiveness of the tine weeder is largely dependent on your speed and structure of the soil (not the angle of the tines). More often than not, it is better to work at slower speeds and make two passes through the soil. Going the opposite direction the second time through.

Basic Settings

Working depth should be about 1" (2.5 cm) so the seedbed is left undisturbed. The tine section should be parallel with the ground.

ROD WEEDERS

Rod weeders are agricultural tools that are designed for mechanical weed control, especially in row crops like cereals, vegetables, and legumes. They work by using rotating rods or tines to agitate the soil around crop rows, uprooting or disturbing weeds. These tools play a crucial role in secondary tillage, providing an effective alternative to herbicides, which makes them well-suited for organic and environmentally friendly farming practices. The rod weeder consists of a rotating square or triangular rod, driven by a chain or gear connected to the implement's gauge wheels. As the rod rotates, it lifts weeds out of the soil. These implements are commonly used in dryland farming for fallow weed control, where land is left fallow to restore moisture. They are particularly useful for post-harvest weed management as they help preserve and even improve surface organic matter, though in humid conditions, weeds may re-root.

Figure 3.18: Rod Weeder

Operating Principle of Rod Weeders

Rod weeders work by using spinning rods or tines that make contact with the soil, either disturbing the surface or uprooting shallow-rooted weeds. They are typically designed for shallow soil penetration and are used primarily to target young, emerging weeds that are still in the seedling stage.

The rods or tines are often mounted on a frame or implement that can be adjusted in height, width, or depth to match the crop's growth stage and row spacing. As the machine moves through the field, the rotating rods agitate the soil, killing or dislodging weeds without damaging the crops.

Varieties of Rod Weeders

Rotary Rod Weeders

- **Description:** These weeding tools feature multiple rods or tines that rotate around a central axis. The rotating motion agitates the soil, cutting or uprooting weeds as the tool moves through the field.
- **Best For:** Use in row crops like corn, beans, and vegetables, especially when weeds are still small and the crop is well-established.
- **Advantages:** Rotary rod weeders are effective at cutting and uprooting weeds in loose or loamy soil. They are particularly good at covering large areas quickly.

Spring Tine Rod Weeders

- **Description:** These use spring-loaded rods or tines that flex and adjust to the contours of the soil, disturbing the surface and uprooting shallow weeds.
- **Best For:** These weeders are commonly used in row crops where the plants have strong enough root systems to withstand minor soil disruption.
- **Advantages:** Spring tine rod weeders are often more flexible and can handle uneven or stony ground better than rigid tine or disc-based weeders. They are well-suited for crops like tomatoes, potatoes, and carrots.

Rigid Rod Weeders

- **Description:** These weeders use rigid, non-flexible rods or tines that penetrate the soil to remove weeds. Unlike spring-loaded models, they don't flex with the terrain and are better suited for smoother fields.
- **Best For:** Used in flat, uniform soil conditions, they can be employed in early tillage stages to control weeds in crops like corn, wheat, and soybeans.
- **Advantages:** These weeders are durable and effective on weed seeds just emerging or on small weeds between rows.

Main Features and Design Elements

- **Adjustability:** Most rod weeders are adjustable, allowing for changes in the spacing between the rods or the depth of penetration into the soil. This ensures that the weeders can be tailored to different crops and field conditions.
- **Rod Shape and Material:** The rods are often made from high-strength materials like steel or spring steel to withstand the stresses of rotation and soil contact. They may be designed with curved or straight shapes to optimize the weeding action.
- **Frame and Mounting:** Rod weeders are typically attached to the rear of a tractor, and they can be mounted either as standalone implements or combined with other tillage tools for broader secondary tillage operations.

Benefits of Using a Rod Weeder

Efficient Weed Control

Rod weeders are effective at uprooting or disturbing weeds in the seedling stage, preventing them from growing larger and competing with crops.

Labor Efficiency

By mechanizing the weeding process, rod weeders reduce the need for manual labor, which can be costly and time-consuming, particularly in larger fields.

Improved Soil Aeration

The action of the rods breaking up the soil helps improve soil aeration, which can benefit plant roots, especially in compacted soils.

Reduction in Herbicide Use

They offer an eco-friendly alternative to chemical herbicides, which is especially beneficial in organic farming and areas with growing concerns about chemical residues in food.

Cost-Effective

Rod weeders are generally less expensive than more complex machines like harvesters or flame weeders, and they require less fuel or power to operate.

Prevents Weed Seed Set

They can uproot weeds before they set seed, reducing the weed seed bank in the soil and preventing future weed infestations.

Drawbacks of Rod Weeders

Row Spacing

These tools are designed to work in row crops. They may not be as effective for crops that are planted more closely together or for densely packed groundcovers.

Weed Growth Stage

They are most effective on young, shallow-rooted weeds. As weeds mature, their roots become harder to uproot, and rod weeders may become less effective.

Soil Conditions

Rod weeders work best in loose or well-tilled soil. In heavy, wet, or very compacted soil, the weeders might struggle to penetrate or could damage the crops.

Maintenance

Regular maintenance is required to keep the rods or tines in good condition, as they are subject to wear from constant soil contact and can be damaged by rocks or debris in the field.

Optimal Applications of Rod Weeders

Early Weed Control

Rod weeders are particularly useful in the early growth stages of crops to control small weeds that have not yet become established.

Post-Emergence Weeding

They can be used post-emergence, after the crop has sprouted but before it has grown too large, allowing farmers to control weeds without harming the crop.

Vegetable and Row Crop Fields

They are ideal for row crops such as corn, soybeans, tomatoes, potatoes, carrots, and lettuce, where weeds are primarily found in between the rows rather than under the crop canopy.

FLAME WEEDERS

Flame weeders are mechanical tools that control weeds by applying intense heat to burn or shock them. This method is mostly used in organic farming and sustainable agricultural practices as an alternative to chemical herbicides. Flame weeding is most effective on young, tender weeds that have not fully established and works best under certain environmental conditions.

The process involves exposing weeds to a propane or butane flame at temperatures between 1,500°F and 1,800°F. This heat damages the cell membranes of the weeds, causing rapid dehydration. A set of burners can cover a large area to eliminate weeds before crop planting, prior to crop emergence, or even to defoliate plants before harvest. Flame weeding is generally used to create a true stale seedbed in organic farming. A canopy over the burner enhances its efficiency by trapping the heat and covering the entire bed. Irrigating a few days before planting can help weeds germinate, allowing them to be burned before the crop emerges.

When directed at the row, flame weeding can control in-row weeds in crops with a protected terminal bud, such as corn and leeks, and in crops like cotton and sunflowers that have tough stems. While flame weeding is limited to certain crops after emergence, it can significantly reduce weed competition, especially for those crops where weeds tend to go to seed. However, flaming does not affect weed seeds that have already fallen to the ground, as they are shielded by the surrounding soil.

Shields can be used to protect flame-sensitive crops from direct heat while enabling efficient weed control near the crop row. Similar to those used for close cultivation of small crops, shields can be attached to tractor-mounted flame weeders. When using a shield, the flame should be directed backward rather than toward the row, as is done when weeding flame-tolerant crops. For handheld and push-type flame weeders, attaching a wheel and shield to the flame wand allows precise weeding on small acreages.

Figure 3.19: In-Row Flame Weeder (Note That the Burners are Staggered to Avoid Forcing Flame Up Onto the Crop Canopy)

The duration of weed exposure to the flame is typically adjusted by altering the speed of movement across the field. Slower speeds lead to more effective weed control but also use more propane. The amount of gas needed for 95% control varies depending on the species and size of the weed. For instance, controlling white mustard (similar to wild mustard) in the two- to four-leaf stage requires 1.5 to 2 times more propane per acre than for seedlings in the zero- to two-leaf stage. Weeds like common lambsquarters, common chickweed, and common groundsel are well controlled with less than 45 pounds of propane per acre when young. However, grasses and broadleaf species with protected buds, such as corn chamomile and common purslane, may regrow after flaming. These species may require multiple treatments or larger doses of propane.

Flame weeding is most effective when the plant surface is dry. If the surface is moist, the heat may be absorbed by the vaporization of moisture, reducing its effectiveness. While soil moisture is not crucial, the soil surface should be smooth to avoid sheltering weeds or reflecting heat onto sensitive parts of the crop during in-row flaming.

Flame weeding carries the risk of starting a fire. Extra caution should be taken to avoid igniting dry grass or brush when turning near field ends, as crop residue may catch fire, even in a clean field. Such fires usually burn out quickly without damaging crops but can spread under windy conditions to ignite surrounding materials. Special attention is required in small-scale vegetable systems where dry crop residue or mulched beds may be near the flamed area. Other heat-based weed control methods, such as using liquid nitrogen or carbon dioxide snow to freeze weeds or infrared radiation to cook them, have been explored. However, these methods require more energy than flame weeding and have not been commercially developed due to their higher cost.

Working of Flame Weeders

Flame weeders apply a controlled flame or heat to weeds, which damages the plant cells, causing them to die. The intense heat from the flame typically doesn't kill the weed immediately but disrupts its cellular structure, causing dehydration, wilting, and eventual death. For perennial weeds or weeds with established root systems, the method might not kill the roots, but it can suppress the growth of above-ground parts.

There are usually two primary approaches:

- **Direct Flame Contact:** This involves directly exposing weeds to the flame, burning the plants' tops and leaves.
- **Radiant Heat:** Some flame weeders use radiant heat (infrared) that doesn't directly touch the weeds but still affects them by raising their internal temperature to a lethal level.

Varieties of Flame Weeders

Hand-Held Flame Weeders

- **Description:** These are portable, typically small devices that can be manually operated. They often resemble large torches or wands, which farmers can carry through the field to target weeds.
- **Fuel Source:** They usually run on propane gas or butane and have a nozzle that directs the flame toward the weed.
- **Best For:** Small-scale applications, garden beds, or organic farming. They're effective in narrow rows of crops or small plots where precision is needed.

Mertis

- Low-cost entry for organic farmers.
- Flexible use in weeding around individual plants or small spaces.
- Easy to maneuver and lightweight.

Tractor-Mounted Flame Weeders

- **Description:** These are larger, more powerful flame weeders that are mounted on tractors or other vehicles. They typically feature multiple burners or nozzles that allow them to cover a wide area in a short period.
- **Fuel Source:** Usually powered by propane or natural gas, connected to the machine via hoses and tanks.

- **Best For:** Medium to large-scale farms, row crops like corn, soybeans, and vegetables where weeds grow between crop rows.

Merits

- Efficient for large fields or row crops.
- Can be calibrated to treat a larger area quickly with minimal labor.
- More consistent application across the field.

Brief Overview of Weeding Flamers (Walk-Behind and Tow-Behind Units)

- **Description:** These weeders are typically towed behind a tractor or pushed by hand, offering a good balance between scale and control. These units have one or multiple burners or nozzles and can be adjusted for different flame intensity.
- **Fuel Source:** Propane or gas-powered.
- **Best For:** Organic farms with larger plots, such as vineyards, vegetable fields, or in-between rows of crops.

Merits

- Customizable settings for different crops.
- Less effort compared to hand-held units, but still affordable and manageable for medium-sized operations.

Infrared (Radiant Heat) Weeders

- **Description:** These weeding devices use infrared heat (rather than an open flame) to target weeds. The heat waves penetrate the plant's surface and damage its internal tissues, eventually causing it to wither and die.
- **Fuel Source:** Typically use propane or electricity for generating infrared heat.
- **Best For:** Suitable for row crops, large-scale vegetable farms, and fields where precision is important.

Merits

- Provides more uniform heat distribution than open flames.
- Can target weeds in a very controlled manner.
- Safer than traditional flame weeding as it doesn't involve an open flame.

Benefits of Flame Weeding

Chemical-Free Weed Control

Flame weeding offers an environmentally friendly and chemical-free alternative to traditional herbicides, which is particularly beneficial in organic farming.

Pre-Emergence and Post-Emergence Control

- Flame weeding is effective both for pre-emergence (before the weeds sprout) and post-emergence (after the weeds have emerged) weed control.
- It is especially useful for young, tender weeds in the seedling stage.

Reduced Risk of Resistance

Since flame weeding is a physical method of control rather than chemical, it helps to avoid the development of herbicide-resistant weeds.

Selective Weed Control

Flame weeders can be highly precise and selectively target weeds between crop rows without harming the crops, provided they are used correctly. This is especially important in fields with delicate crops like vegetables or herbs.

Minimal Soil Disruption

Unlike tilling or mechanical weeding, which can disrupt the soil and potentially harm beneficial organisms, flame weeding involves no soil disturbance, making it ideal for maintaining soil health.

Speed and Efficiency

Flame weeding can cover large areas quickly, especially with tractor-mounted systems. This is particularly useful for row crops and large fields where labor-intensive methods like hand weeding would be impractical.

Drawbacks of Flame Weeding

Limited Effectiveness on Perennial Weeds

Flame weeding usually kills only the above-ground portion of the plant. Perennial weeds with deep root systems (like dandelions or perennial grasses) might survive the treatment and regenerate.

Weather and Environmental Conditions

Flame weeding is most effective on dry, warm days when weeds are small and tender. In humid, wet, or windy conditions, the effectiveness can be reduced, and there's a risk of fire spreading in dry, windy conditions.

Energy and Fuel Costs

While propane and butane are efficient fuel sources, they still add costs to flame weeding, especially for larger-scale operations. The environmental impact of burning fossil fuels for weed control is also a concern for some farmers.

Risk of Crop Damage

If not used carefully, flame weeding can damage delicate crops or young plants. Farmers must ensure the correct flame intensity and distance to avoid harming the crops.

Labor Intensive

Although tractor-mounted systems cover large areas, hand-held flame weeders can be labor-intensive, especially in large fields or dense weed populations.

Limited Effect on Deep-Rooted Weeds

Since flame weeding targets the above-ground parts of weeds, it does not affect the roots. Deep-rooted weeds like thistles or perennials might survive the treatment and regrow.

Optimal Applications of Flame Weeding

Organic Farming

Flame weeding is an essential tool in organic farming, where the use of chemical herbicides is prohibited. It provides an effective way to manage weeds without synthetic chemicals.

Pre-emergence Weed Control

Flame weeding is often used as a pre-emergence technique to kill weed seeds before they germinate or shortly after they emerge from the soil.

Vegetable and Row Crops

Flame weeders are ideal for controlling weeds in crops like tomatoes, lettuce, carrots, corn, and potatoes, where weeds tend to emerge between the crop rows.

Vineyards and Orchards

These tools are also effective in vineyards and orchards, where row spacing is wide enough to accommodate the flames, and chemical herbicides may not be desirable.

STEAM AND HOT WATER WEEDERS

Steam weed control uses hot water to vaporize weeds, making it an effective method for eliminating weeds along footpaths, in public spaces such as schools and parks, as well as in agricultural environments. This approach is typically chemical-free, relying solely on water to treat the weeds.

Like flame weeders, steam and hot water weeders kill by disrupting plant surfaces, leading to subsequent desiccation.

Steam and hot water weeders function similarly to flame weeders by damaging the plant surfaces, causing them to dry out. The benefit of steam and hot water weeders over flame weeders is that the flame is contained within a boiler, reducing the risk of starting a fire. However, they consume significantly more energy per acre than typical flame weeders. Hot water weeders are more energy-efficient than steam weeders as they do not need to bring water to a boil, whereas steam weeders require less water. Although these methods are gaining some popularity in urban weed management, their adoption in agriculture has been limited due to high fuel consumption and the potential for soil compaction caused by transporting large amounts of water over tilled fields.

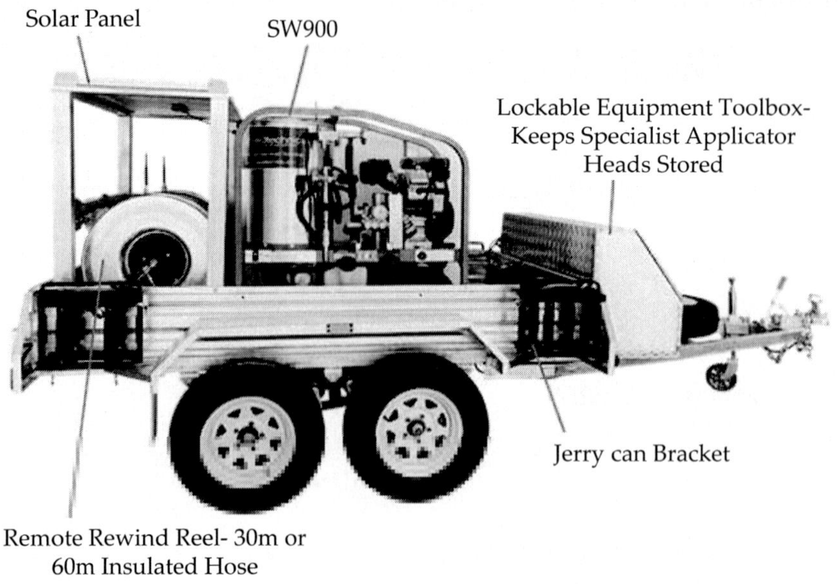

Figure 3.20: Steam Weeders

Working of A Steam Weeder

Steam weed control operates on a similar principle to hot foam herbicide alternatives like Foamstream. Essentially, it generates temperatures close to boiling to eliminate the cells in the above-ground parts of the weed. As a result, the plant dies while the root remains alive, much like with herbicide treatments. Repeated treatments are necessary to weaken the plant further until the roots eventually die. The pressurized steam continuously shrivels the plant, preventing it from processing nutrients or photosynthesizing.

Variants of Steam and Hot Water Weeders

Tractor-Mounted Steam or Hot Water Weeders

- **Description:** These systems are mounted on tractors and are used to treat large areas or row crops. They have tanks or boilers to generate steam or hot water, which is distributed through a network of hoses or pipes to nozzles that spray the steam or water on the weeds.
- **Best For:** Large farms, row crops, or vineyards where weed control is needed in between crop rows.

Merits

- Ideal for large-scale operations.
- Can treat vast areas in a short time.
- Reduces the need for herbicides in large fields.

Walk-Behind or Handheld Steam/Hot Water Weeders

- **Description:** These are smaller, portable machines designed for use in garden beds, greenhouses, or small farms. They often consist of a portable boiler or tank that generates steam or hot water, and a handheld nozzle directs it to the weeds.
- **Best For:** Smaller-scale operations, gardens, or areas with narrow row spacing.

Merits

- More flexible and easier to maneuver in tight spaces.
- Suitable for organic farming and garden maintenance.
- Portable and relatively easy to use.

Trailer-Mounted Steam/Hot Water Units

- **Description:** These units are typically towed behind a tractor and are equipped with a larger capacity for steam or hot water production. They are designed for larger fields or extensive areas that require more consistent, high-volume treatment.
- **Best For:** Large-scale farms and fields where high-volume, continuous weeding is required.

Merits

- High-efficiency for treating large plots of land.
- Can operate for extended periods, covering more ground with fewer stops for refueling.

Advantages of Steam and Hot Water Weeding

Chemical-Free Weed Control

Steam and hot water weeding provide a natural, chemical-free solution to weed control, which is ideal for organic farming and reducing reliance on herbicides.

Environmentally Friendly

Since they don't use herbicides or other chemical pesticides, steam and hot water weeders are a more eco-friendly method for managing weeds. They don't leave harmful residues in the soil or on crops, making them ideal for sustainable farming.

No Soil Disturbance

Unlike tillage-based methods, steam and hot water weeders do not disturb the soil. This helps preserve the soil structure and the beneficial organisms living in the soil, such as earthworms and microbes.

Precise Application

The application of heat is highly targeted, which means it can be applied specifically to the weeds, reducing the risk of damaging nearby crops.

Hot water and steam weeding systems allow for adjustable nozzles or other precision tools to control the intensity and coverage, ensuring effective treatment with minimal waste.

Effective for Early-Stage Weeds

These methods are particularly effective against young weeds and seedlings. Steam and hot water can kill weeds at various stages of growth, making them a valuable tool for pre-emergence or early post-emergence weed control.

Reduces Herbicide Resistance

By using thermal methods for weed control, farmers can reduce reliance on chemical herbicides, thereby minimizing the risk of developing herbicide-resistant weeds.

Works on Both Annual and Perennial Weeds

These methods can be effective for managing both annual weeds (like dandelions, crabgrass, etc.) and some perennial weeds (like bindweed), particularly when applied early in the growing season.

Drawbacks of Steam and Hot Water Weeding

Effectiveness on Mature Weeds

Steam and hot water weeders are more effective on young, tender weeds. As weeds mature and their root systems become more established, it becomes harder to completely kill them with thermal methods, especially for perennial weeds.

Energy and Fuel Consumption

- These systems require significant amounts of energy to generate the steam or heat, often in the form of propane or electricity, which can add operational costs.
- Large-scale systems, especially those using propane, can incur high fuel costs.

Weather Conditions

Thermal weeding is more effective in dry, sunny conditions. Rain or excessive humidity can hinder the effectiveness of steam and hot water applications, as the moisture can reduce the impact of heat on the weeds.

Risk of Overheating or Crop Damage

- If not carefully managed, steam or hot water can damage crops or cause harm to surrounding soil and beneficial organisms.
- The heat needs to be applied carefully and precisely to avoid damaging the desired crops.

Labor Intensity

Some systems, mainly handheld steam weeders, may require more manual effort and be more labor-intensive compared to other methods like herbicide spraying or mechanical weeding.

Optimal Uses of Steam and Hot Water Weeding

Organic Farming

These methods are especially valuable for organic farmers who cannot use chemical herbicides to manage weeds. Steam and hot water provide an effective, natural solution to keep fields and gardens free from weeds.

Greenhouses and Gardens

Hot water and steam weeding systems are useful in smaller-scale applications such as greenhouses, garden beds, and around individual plants, where precise targeting is crucial.

Perennial Weed Control

These methods are especially effective for managing perennial weeds when combined with other management strategies (like mulching or crop rotation).

Vineyards and Orchards

Ideal for use in vineyards and orchards, where herbicides are often avoided to maintain soil health and meet organic certification standards.

Vegetable and Row Crop Fields

Steam and hot water weeders are ideal for control of weeds in row crops (such as corn, potatoes, and tomatoes) where weeds tend to emerge in between crop rows.

ELECTRICAL DISCHARGE WEEDERS

Electrical Discharge Weeders are an innovative and environmentally friendly method of weed control that utilizes electricity to kill or damage weeds. This technology is especially useful for those who want to avoid the use of herbicides or mechanical methods. The principle behind electrical discharge weeding involves using high-voltage electrical pulses to target and destroy weed plants.

Working of Electrical Dischaege Weeders

Electrical discharge weeders use high-voltage electrical currents to damage or kill weeds, specifically by targeting the conductive nature of plants. Below mentioned are some of the key points to understand the working of an electrical discharge weeders.

Electrical Pulse Application

The weeder is equipped with electrodes or conductors, which are applied to the plants or soil. These electrodes generate an electrical field that is directed toward the weed. When the electrical current passes through the plant, it disrupts the plant's internal processes, causing cell membranes to break down and tissues to die. The current primarily targets the vascular system of the plant (the system that carries water and nutrients), causing the plant to wilt, dry out, and die.

Types of Electrical Discharge Weeders

- **Direct Current (DC) Systems:** Use direct current to pass through the plant. It causes localized heating and tissue damage, leading to plant death.
- **Pulsed Electrical Fields (PEF):** Use short, high-voltage pulses to shock and disrupt the plant's cellular structure without the need for continuous electricity. This pulsing is particularly effective at disrupting the plant's physiological processes.
- **Electrochemical Methods:** Sometimes combined with a conductive liquid (such as water or a specialized solution), these systems use electrical discharge to break down the chemical bonds in the plant tissues.

Components of Electrical Discharge Weed Control Systems

Power Supply

The weeder is powered by an external source of electricity, typically batteries or a generator. The power supply provides the high voltage required for the electrical pulses.

Electrodes or Electrodes Plates

These conductors are placed in contact with the weeds. They can be mounted on walking units, tractors, or robots designed to cover larger areas.

Controller

The system is equipped with a controller that manages the voltage output, pulse timing, and duration of the electrical discharge. It allows operators to adjust settings based on the type of weed, plant size, and growth stage.

Power Conversion Mechanism

In some systems, the electrical current may need to be converted or modified (for example, from alternating current to direct current) to be suitable for weed control.

Advantages of Using Electrical Discharge Weeders

Chemical-Free Weed Control

Electrical discharge weeders provide a chemical-free method of weed control, making them ideal for organic farming or eco-friendly agricultural practices.

Reduced Herbicide Use

By replacing or reducing the need for chemical herbicides, this method helps prevent the development of herbicide-resistant weeds and minimizes environmental pollution.

Safety

Compared to flame or mechanical weeding methods, electrical discharge weeders are generally safer in terms of fire risk, especially in dry conditions, as they don't use open flames.

No Soil Disturbance

Since it works above ground and does not require tillage, there is no soil disturbance, which preserves the soil structure and prevents the destruction of beneficial organisms in the soil.

Precision

Electrical discharge weeders can be designed for precision application. The system can target specific weeds without affecting nearby crops, making it suitable for row crops or garden beds.

Sustainable

This method is energy-efficient, especially in smaller-scale operations. It also doesn't leave harmful residues in the soil or on crops, making it a sustainable option for long-term weed management.

Drawbacks of Electrical Discharge Weeders

Effectiveness on Large & Mature Weeds

Electrical discharge weeders are most effective on young, tender weeds and may not be as effective on mature, deep-rooted weeds. Perennial weeds with well-established root systems may survive the electrical treatment.

Energy Consumption

The process of generating high-voltage electrical pulses requires energy, and in large-scale operations, this could lead to high electricity consumption. For large farms, this could also mean additional costs for electricity or fuel for power generation.

High Initial Cost

The equipment for electrical discharge weeding, including power supplies, electrodes, and controllers, can be expensive to purchase and maintain. The initial investment may be a barrier for smaller-scale farmers.

Weed Size and Density

The system may be less effective on weeds that are too small or too dense. Achieving uniform coverage can be challenging, and the weed control may be less effective in heavily infested areas.

Limited Availability and Development

Electrical discharge weeders are still relatively new in the agricultural sector and may not be as widely available or fully optimized for various types of crops or geographical areas.

Uses of Electrical Discharge Weeding

Row Crops and Vegetable Farms

Electrical discharge weeders are suitable for row crops, vegetable fields, and areas with narrow spacing where weeds tend to grow between crops. They offer precise control, allowing the farmer to target weeds without destroying the crops.

Greenhouses

Electrical discharge weeders are also useful in greenhouses where precision is required. Since greenhouses are controlled environments, these weeders can be used to maintain weed-free conditions without the need for herbicides.

Vineyards and Orchards

These weeders can be effective in vineyards and orchards, where herbicide use is often limited due to the risk of contamination, and there is a need to manage weeds around the vines or trees without damaging the crops.

Organic Farming

Since electrical discharge weeders don't rely on chemicals, they are highly valuable in organic farming where chemical herbicides are prohibited or undesirable. They provide an eco-friendly solution for weed management.

Smaller Farms or Garden Beds

For smaller-scale farming or home gardens, these weeders offer a safe, non-toxic way of controlling weeds, especially in areas where it's impractical to use heavy equipment or chemicals.

Final Thoughts

Through discussions with various agricultural equipment dealers in Bangalore, it was noted that agricultural weeders are essential tools for those engaged in crop production, gardening, landscaping, or environmental conservation. They can minimize the reliance on herbicides, enhance crop yields, lower labor costs, and boost efficiency. Given the wide range of agricultural weeders available, selecting the appropriate tool for the task is crucial to achieving optimal results.

4

Rotavator

A rotavator, also known as a Rotary Tiller, is a tractor-powered tillage tool used for oil cultivation. It works by cutting, mixing, and pulverizing the soil through a series of rotating blades. This tool is an affordable and efficient option for seedbed preparation and can replace equipment like disc harrows, cultivators, and levelers.

The rotavator features spinning blades that break through the soil, powered by the tractor's PTO. It increases soil structure by breaking up clods, improving water absorption, and eliminating weeds and crop residue. Compared to a plough, a rotavator allows farmers to prepare land for sowing in fewer passes. As a powered machine, it ensures optimal soil conditions while helping to control pests and prevent crop diseases.

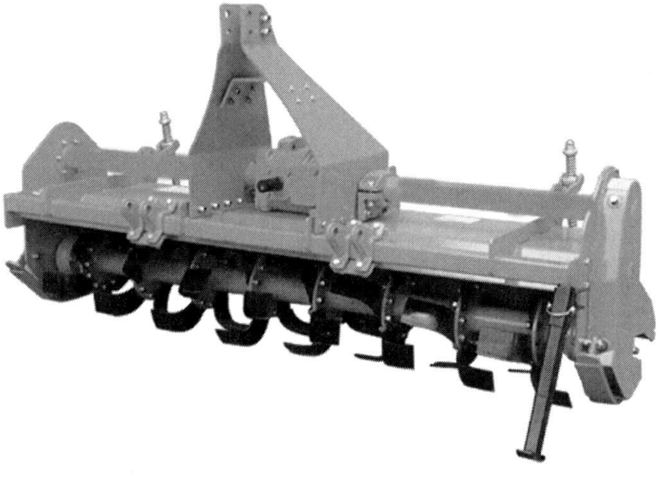

Figure 4.1: Rotavator Featuring Spinning Blades

A rotavator or a rotary tiller is one of the most efficient tools used in modern-day agriculture. It uses rotating blades and advanced machinery to pulverize the soil. According to several case studies, rotavators are considered one of the best economical and effective tools for soil bed preparation.

DESIGN FEATURES OF A ROTAVATOR

The inter-row gangs, mounted with parallel linkage, are individually suspended and feature adjustable heads ranging from 5 to 60 inches in width, depending on knife width and arrangement. These gangs are attached to a main toolbar, with a chain drive powering the heads via an elevated hex-shaped PTO drive shaft positioned parallel to and slightly below the toolbar. Heavy-duty models come equipped with cast-iron gearboxes for added durability. Some manufacturers offer two chain housing lengths to better accommodate raised bed applications. For in-row weed control, additional furrower attachments are required to direct soil into the rows.

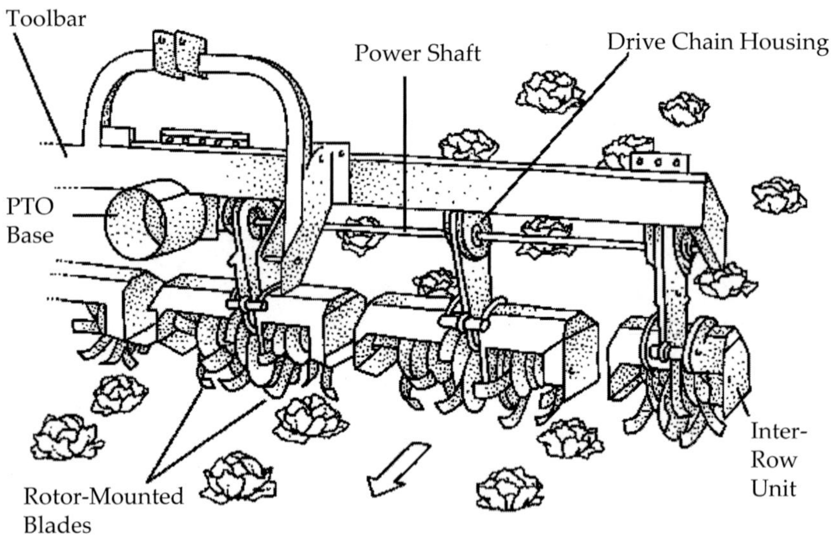

Figure 4.2: Rotavator

USAGE IF ROTAVATOR

The rotary tiller consists of a series of blades powered by the tractor PTO, which effectively breaks through, mixes, and pulverizes the soil. Its frame can be adjusted to control the tilling depth, transforming the soil from a hard surface to a finer structure for improved seed sowing and planting. This affordable tool offers several advantages:

- It reduces soil thickness, breaks lumps, and removes clods, ensuring better seed germination.
- It clears out previous crop residue and stubble.

- Increased soil aeration promotes water infiltration, nutrient absorption, moisture conservation, and root penetration.
- It prevents weed seed growth.
- By improving soil structure and stability, it enhances crop growth potential while reducing soil erosion risks.

Functions of a Rotavator

- Breaking up and leveling soil
- Seedbed preparation
- Suitable for both primary and secondary tillage
- Saves time and minimizes fuel consumption
- Prepares soil efficiently
- Removes crop residue from previous cultivation
- Versatile models for different tasks
- Adjustable frame for desired depth
- Completely eradicates weeds
- Capable of tilling large areas in shorter timeframes
- Provides better soil pulverization compared to other tillage tools
- Loosens and aerates soil up to a depth of 10–15 cm

VARIANTS OF ROTAVATOR

Ground Tiller

A ground tiller, also called a rotavator, breaks up soil using rotating blades. It loosens and mixes the soil, improving drainage and irrigation, making it highly effective for flower beds.

Heavy Duty Rotavator

These are the most powerful rotavators, suited for preparing large tracts of farmland. They are especially beneficial for challenging terrains and are often preferred for heavy-duty agricultural tasks.

Medium Duty Rotavator

With more power than a ground tiller, medium-duty rotavators are ideal for gardens and allotments. They are durable, provide superior finishes, and limits manual effort while preparing fields.

TRACTOR ROTAVATOR ASSEMBLY

Agricultural rotavators are Crucial tools for preparing soil, ensuring effective crop growth. They pulverize, mix, and level soil for crops like vineyards and orchards. Powered by a PTO shaft, these tools are highly efficient for both farming and gardening. For example, Mitra Rotomaster, a popular tractor rotavator, offers multipurpose capabilities.

Key Parts Include

- **Blades:** 20-24 Boron Steel L-Type blades with a helical arrangement for maximum efficiency.
- **Single-Speed Gearbox:** Enables primary and secondary tillage.
- **Adjustable Bottom Skid:** Reduces maintenance costs.
- **Safety Device (Shear Bolt):** Protects equipment from high torques and vibrations.
- **Rotary Shaft:** Enhances Functionality.

Merits

- Reduced tractor load due to helical blades, ensuring ease of use.
- Heavy-duty gear drives for low maintenance and extended durability.
- High working depth (up to 5 inches) for effective land preparation.
- Saves farmers' time and effort compared to traditional land preparation methods.

Alternate Uses of a Rotary Tiller

A rotary tiller is versatile equipment used for both primary and secondary tillage in farming and gardening. Self-propelled models are ideal for domestic gardens, while tractor-mounted versions are used on farms for breaking down soil and preparing land for new cultivation. Other applications include cleaning oilfield pipes and preparing roads.

Prepping Your Garden

Rotary tillers break clods and prepare soil for planting. In heavy vegetation, a single pass followed by rainfall enhances results. Primary tillage requires slow speeds, while secondary tillage allows higher gears for efficiency.

Cultivating Your Garden

Rotary tillers are excellent for crop cultivation. Operating at a shallow depth avoids damaging roots, while higher speeds ensure effective tilling.

Growing and Incorporating Organic Matter

Rotary tillers are excellent for incorporating cover crops and green manure, which enhance the health of garden soil. They also help improve soil structure and tilth by mixing in vegetation and compost. After preparing the seedbed, you can broadcast seeds and use the tiller at high speed and shallow depth to cover them. This method works particularly well for larger seeds, such as peas. However, for smaller seeds like ryegrass, a tiller may cover them too deeply, making raking or dragging a more effective approach.

Middlebusting

Wide sweeps, also known as middlebusters, are the most useful accessory for your tiller. A middlebuster attaches to the rear of the tiller, allowing you to dig a more shallow furrow. Two passes with a middlebuster throws up a raised bed, which can be important in certain areas. Then, you can level the top of that raised bed to plant your crops.

Trenching

Tillers are versatile tools with applications beyond gardening, such as digging ditches or trenches. To begin, run the tiller along the desired ditch line to loosen the soil. Afterward, shovel out the loosened soil and repeat the process until the trench reaches the desired depth.

UPKEEP AND SERVICING OF TRACTOR ROTAVATOR

Tractor rotavators are widely used for various vineyard and orchard crops, often requiring frequent use. Regular inspections of the rotary shaft and tiller engine are essential to ensure proper functionality. Since the rotavator is heavily used for soil preparation, the tillers and shaft must be replaced after every 40 to 50 hours of operation.

Applications of Tractor Rotavator

- When operating a tractor rotavator or similar agricultural equipment, safety precautions are crucial.
- Only trained individuals should operate the rotavator.
- Keep children away from the equipment.
- Avoid loose-fitting clothing to prevent the risk of entanglement in moving parts.
- Never touch the blades or rotary tillers while they are running.
- Ensure all gearboxes, shafts, and blades are securely attached before starting the machine.

Benefits of Using a Rotavator

Rotavators simplify soil preparation, enhancing nutrient distribution to plants and improving soil structure, which boosts crop yields and profitability. Most industrial rotavators are self-propelled, offering adjustable speeds via gearboxes while maintaining consistent blade rotation. Their ease of operation and powerful engines make them efficient for tilling large areas. These tools improve farming efficiency and support better crop growth in well-prepared soil.

Key Considerations for Using a Rotary Tiller

- **Attachments:** Tillers can be equipped with attachments like dethatching tines, rotary brushes, snow/dirt blades, and even generators, making them versatile for various tasks beyond gardening.
- **Adjusting the Working Depth:** The depth can be modified using skid shoes, with larger tillers offering deeper cutting. For large gardens, keep the depth under 6 inches for optimal results.
- **Considering Working Width:** When purchasing a tiller, ensuring wither its width matches or exceeds the tractor's rear tire base is essential. Overlapping each pass while tilling helps ensure no part of the garden is left untended.

Using a Rotary Tiller in a New Garden

When establishing a new garden, it is advisable to plow the soil during the fall and allow it to rest over the winter months. In the spring, use a rotary tiller to work through the soil multiple times until it is tilled 4 to 6 inches deep, with no large clumps remaining. Operators should start slowly and avoid going too deep initially, as tilling too quickly may leave areas of soil untouched.

Adjusting the Tailgate

The tailgate on rotary tillers can be adjusted to suit the desired soil texture. A more open tailgate allows larger soil clods to pass through, resulting in a coarser soil texture being left behind.

Benefits of Using a Rotavator/Rotary Tiller

Soil Preparation and Tilling

A rotavator is effective for both primary and secondary tillage activities, providing uniform soil preparation for seedbeds. The rotary tiller's blades break down soil into fine particles and eliminate clods, creating a finely leveled surface ideal for seed germination. Additionally, its sharp blades can penetrate hard and stony soils, making it suitable for various soil types.

Weed Control

Rotavators are instrumental in weed control, as their rotating blades uproot and bury weeds during tillage or seedbed preparation. By cutting parasitic plants into smaller pieces, the likelihood of regrowth is significantly reduced. This not only clears the field but also enriches the soil with organic matter, benefiting germinating seeds.

Incorporation of Crop Residues

Relay cropping, commonly practiced in India, often leaves behind a high volume of crop residues. Rotavators effectively incorporate these residues into the soil during primary tillage, grinding them into smaller pieces and blending them with the soil. This process enhances the organic matter and nutrient content of the soil, promoting healthy plant growth.

Mixing Soil Amendments

Rotavators facilitate the uniform mixing of fertilizers, pesticides, and natural humus throughout the field. This ensures a balanced nutrient distribution, supports healthy root development, and optimizes input costs. By reducing wastage and increasing efficiency, rotavators contribute to higher crop yields and improved quality.

Aeration of Soil

Proper soil aeration is essential for smooth seed germination and plant growth. A rotavator breaks soil into evenly fine particles, creating air pockets that store oxygen and other essential gases. These air pockets create a favorable environment for germinating seeds and plants, ensuring better crop growth and development.

Seedbed Preparation

Rotavators are highly effective for seedbed preparation during secondary tillage. They offer versatility and adaptability for creating suitable seedbeds in various soil conditions and for different crop types. Compared to other modern agricultural tools, rotavators have been shown to achieve the highest seed germination rates.

Pudding for Paddy Fields

Certain crops, such as rice, require a process called puddling, which involves creating waterlogged barriers to keep plants submerged. Rotavators are usually used to establish puddling systems in fields, ensuring consistent moisture levels and reducing water filtration rates. This technique enhances plant growth, maintains uniform water distribution, and levels the field effectively.

Prevention of Soil Erosion

Unlike traditional plowing methods, which can expose soil to erosion by wind and water, rotavators help mitigate soil erosion. They maintain a protective layer of mulch or crop residue on the surface, shielding the soil from raindrop impact and wind. Improved soil structure from rotavating also enhances water infiltration, reduces runoff, and preserves nutrient-rich topsoil.

Covering Multiple Tasks

Rotavators perform multiple tasks such as plowing, harrowing, and leveling in a single pass, significantly reducing labor and time required for soil preparation. Their multitasking capability eliminates the need for additional machines and minimizes operational costs. Designed for efficiency, rotavators cover large areas quickly, making them ideal for medium to large-scale farming operations.

5

Sowing Equipment

Sowing equipment plays an important role in modern agriculture by ensuring efficient and precise planting of seeds, which is fundamental for achieving optimal crop yield and quality. The evolution of agricultural machinery has transformed traditional methods of sowing, replacing labor-intensive practices with mechanized solutions that save time, reduce human effort, and enhance productivity. These tools and machines are designed to address the diverse requirements of various crops, soil types, and climatic conditions, making them indispensable for farmers seeking sustainable farming practices.

The key objective of sowing equipment is to place seeds at the desired depth and spacing, ensuring uniformity in the field. This precision not only promotes better germination rates but also facilitates proper growth and development of crops. Modern sowing machines are equipped with advanced features such as adjustable depth controls, seed metering systems, and compatibility with fertilizers, enabling farmers to customize their planting process according to specific needs. Additionally, the integration of technology in these machines has further enhanced their efficiency, with GPS-guided systems and automation ensuring minimal wastage and maximum accuracy.

Sowing equipment also contributes to soil conservation by minimizing disturbance during the planting process. Techniques such as zero-tillage and minimum-tillage sowing are made possible with specialized machinery, reducing erosion and preserving soil health. Furthermore, these tools are designed to work seamlessly in various terrains and conditions, ensuring consistent performance in diverse agricultural landscapes. By optimizing resource utilization and improving planting accuracy, sowing equipment supports farmers in meeting the growing demand for food while adhering to environmentally responsible practices. The advancements in this field continue to revolutionize agriculture, offering innovative solutions that align with the goals of efficiency, sustainability, and profitability.

METHODS OF SOWING

There are various methods of sowing, including broadcasting, dibbling, drilling, seed dropping behind the plough, transplanting, hill dropping, and check row planting. Each method is suited for specific crops and soil conditions, ensuring efficient seed placement and optimal germination.

Broadcasting

Broadcasting involves scattering seeds randomly over the surface of seedbeds. This method can be performed manually or mechanically. In manual broadcasting, the uniformity of seed distribution depends immensely on the skill of the person performing the task. Once seeds are scattered, they are typically covered using a plank or other equipment. This method typically requires a higher seed rate. For large-scale operations, mechanical broadcasters are used. These machines distribute seeds evenly across the seedbed at controlled rates, improving efficiency and reducing labor.

Dibbling

Dibbling refers to placing seeds in individual holes made in the seedbed and then covering them with soil. This method ensures seeds are placed at a specific depth and spacing. A specialized tool called a dibbler is used for this process. The dibbler is generally a conical-shaped instrument designed to create uniform holes in the soil. Some hand dibblers come equipped with multiple conical projections in a frame, allowing several holes to be made simultaneously. While this method is precise and particularly suited for crops like vegetables, it is time-consuming and not ideal for small seeds. As a result, dibbling is often limited to specific crops where precision planting is essential.

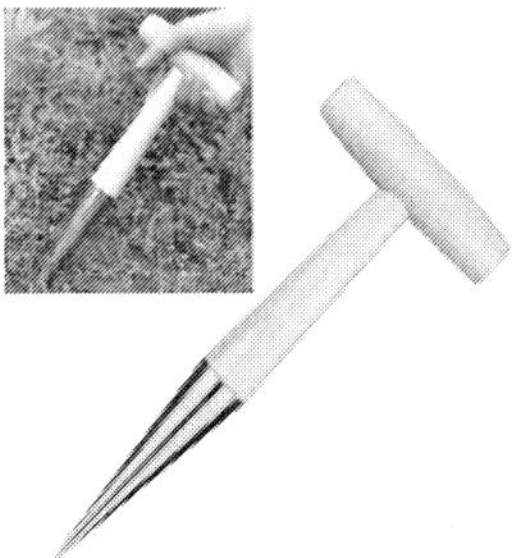

Figure 5.1: Dibbler

Seed Dropping Behind the Plough

This traditional sowing method is widely practiced by farmers in rural areas. It is usually used for crops like maize, gram, peas, wheat, and barley. A person walks behind a plough, dropping seeds into the furrows created by it. This process can be facilitated using a device known as a malobansa, which consists of a bamboo tube with a funnel-shaped mouth attached to the plough handle. One person feeds seeds into the funnel, while another operates the plough and directs the bullocks. This method is quite effective but slow and labor intensive.

Drilling

Drilling involves sowing seeds in continuous furrows and covering them with soil. This method ensures proper depth, spacing, and quantity of seeds for optimal growth. Seed metering can be done manually or mechanically, and the number of rows planted may vary. Drilling can be performed using three primary techniques: sowing behind the plough, bullock-drawn seed drills, or tractor-drawn seed drills. This method is highly efficient in achieving uniformity and precise seed placement.

Transplanting

Transplanting involves growing seedlings in a nursery bed before planting them in the main field. This method is commonly used for crops such as paddy, vegetables, and flowers. Although it is a time-intensive process, transplanting allows for the careful placement of seedlings to maximize growth potential. Specialized equipment called transplanters can be used to facilitate this operation.

Hill Dropping

In this method, a small cluster of seeds is dropped at specific intervals rather than in a continuous flow. The spacing between each hill in a row remains consistent. This method is particularly effective for crops requiring fixed spacing. Equipment used for hill dropping is referred to as a planter.

Check Row Planting

Check row planting ensures uniform row-to-row and plant-to-plant spacing. Seeds are planted in straight parallel furrows, with rows intersecting at right angles. This precise method is carried out using a check row planter, which ensures consistency in seed placement and optimal field utilization.

Safety Measures for Sowing Seeds

Proper precautions must be observed during the sowing process to ensure optimal seed germination and plant growth. Firstly, seeds must be sown at an appropriate depth in the soil, ensuring they are not too shallow to be displaced by wind or animals nor too deep to hinder germination. The seeds used should be disease-free to avoid compromising crop health. Adequate spacing between seeds must be maintained to allow each plant access to sufficient light, nutrients, and water. Additionally, care should be taken to ensure that all seeds are sown uniformly, providing equal conditions for growth. Proper depth placement is essential, ensuring seeds are placed where they can receive the necessary moisture and warmth for germination.

SEED DRILL

A seed drill is an essential agricultural device used to sow seeds efficiently by placing them at a precise depth and distance in the soil. This machine ensures uniform seed distribution and consistent depth placement while covering the seeds with soil after sowing. Seed drills can be powered by either tractors or bullocks and are suitable for planting various crops such as maize, wheat, peas, and mustard.

Based on the power source, seed drills can be categorized into (i) bullock-drawn seed drills and (ii) tractor-drawn seed drills. Bullock-drawn seed drills, in turn, are classified into two types depending on the seed metering mechanism: (i) manually metered seed drills and (ii) mechanically metered seed drills. In manually metered seed drills, an individual drops seeds into the furrows, whereas in mechanically metered seed drills, a mechanical device known as the seed metering mechanism regulates seed placement.

Both bullock-drawn and tractor-drawn seed drills come in various designs to cater to specific agricultural requirements. These machines are indispensable for achieving efficient, large-scale sowing operations while maintaining uniform seed placement, which is critical for optimal crop growth and yield.

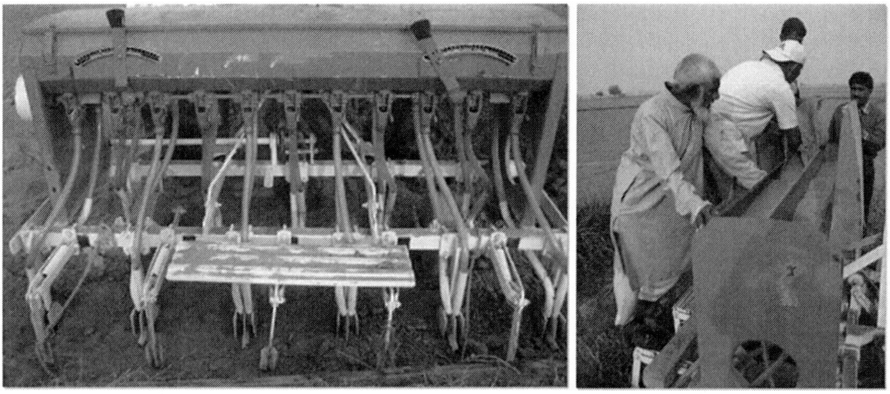

Figure 5.2: Seed Drill for Wheat

ROLES OF A SEED DRILL

Seed drill performs the following roles:

- Carrying the seeds.
- Opening furrows at uniform depths
- Metering the seeds
- Depositing the seeds in furrows in an acceptable pattern
- Covering the seeds and compact the soil around the seed

SEED-CUM-FERTILIZER DRILL

A Seed-cum-Fertilizer Drill is a modern agricultural tool designed to simultaneously plant seeds and apply fertilizers in a single operation. By combining the processes of seed sowing and fertilizing, it enhances efficiency, reducing both time and labor for farmers. This implement ensures that seeds are sown at the correct depth and spacing while applying the appropriate amount of fertilizer to promote healthy crop growth.

Seed drills equipped with fertilizer dispensing attachments are referred to as seed-cum-fertilizer drills. These drills distribute both seeds and fertilizers in a uniform pattern. The seed-cum-fertilizer drill features a large seed box that is divided into two sections: one for seeds and another for fertilizer.

Functions of a Seed-cum-Fertilizer Drill

- To carry seeds and fertilizer in separate compartments.
- To open furrows at consistent depths.
- To meter the distribution of seeds and fertilizers.
- To place seeds and fertilizer in the furrows in a uniform pattern.
- To cover the seeds and fertilizers and compact the soil around the seeds.

Parts of a Seed Drill

A seed drill with a mechanical seed metering device consists of several essential components, including: (i) Frame (ii) Seed Box (iii) Seed Metering Mechanism (iv) Drive Transmission System (v) Furrow Openers (vi) Covering Device (vii) Clutch (viii) Hitch Frame (ix) Transport Wheels.

Frame

The frame is typically constructed from mild steel angle sections and flats. It is designed to be robust enough to handle all types of loads during operation. All other components of the seed drill are mounted on this frame, providing stability and support.

Seed Box

The seed box is a container, usually made of mild steel or galvanized iron, with a lid. In some models, a small agitator is installed at the bottom of the box. This agitator keeps the seeds in motion during operation, preventing them from clogging. The seed metering mechanism is located at the bottom of the seed box, ensuring accurate seed dispensing.

Seed Metering Mechanism

The seed metering mechanism is a vital part of a seed drill or planter, ensuring the accurate and uniform distribution of seeds during sowing. This system regulates the rate and spacing of seed placement, essential for proper crop germination and development. It minimizes seed wastage, enhancing crop productivity. The mechanism, which picks up seeds from the seed box and delivers them into the seed tube, can be of several types:

- Fluted feed type
- Internal double run type
- Cup feed type
- Cell feed type
- Brush feed type
- Auger feed type

- Picker wheel type
- Star wheel type

Usually, the seed metering mechanism is located at the bottom of the seed box.

Drive Transmission System

The drive transmission system consists of a wheel, sprocket-chain assembly, and a driven shaft connected to seed-picking discs. As the seed drill moves through the field, the drive wheel rotates from its contact with the soil, which in turn rotates the sprocket wheel. The connected chain transmits motion to the driven wheel sprocket, which rotates the shaft and activates the seed metering discs.

Furrow Openers

Furrow openers create furrows in the soil for seed placement. Various types of furrow openers are used, including:

- Hoe type
- Shoe type
- Stub runner type
- Full or curved runner type
- Single disc type
- Double disc type

In some cultivator-type seed drills, the tines themselves function as furrow openers.

Covering Device or Furrow Closer

The covering device or furrow closer seals the furrow after the seeds have been placed inside. It ensures that the seeds are covered with soil, promoting better germination. The covering process is typically achieved using chains, bars, packers, rollers, or press wheels in various designs and sizes.

Transport Wheel

The transport wheels are mounted on an axle, facilitating the movement of the drill on roads. Iron wheels are commonly used, although some models feature pneumatic wheels. One of the transport wheels is equipped with an attachment that transmits motion to the seed metering mechanism, enabling its operation during fieldwork.

Operating Principle of a Seed-cum-Fertilizer Drill

Preparation

The machine is loaded with seeds and fertilizers into their respective hoppers. The operator adjusts the seed and fertilizer rate according to the crop and soil type for efficient sowing and fertilization.

Furrow Creation

As the tractor or bullock pulls the drill forward, the furrow openers penetrate the soil, creating furrows at the specified depth and spacing. This ensures proper placement of seeds and fertilizer.

Covering the Furrow

The covering device ensures that the seeds and fertilizers are properly covered with soil, protecting them from exposure to air, preventing evaporation, and creating optimal conditions for germination.

Seed and Fertilizer Distribution

The seed metering mechanism releases seeds into the seed tubes, which direct them into the furrows. Simultaneously, the fertilizer metering mechanism releases fertilizer into separate tubes, placing it either with the seeds in the same furrow or in a nearby band for effective fertilization.

Uniform Application

The depth control mechanism and metering system work together to ensure uniform application of seeds and fertilizers, providing consistent growth conditions for the crops.

VARIANTS OF SEED METERING MECHANISMS

Some of the key seed metering mechanisms used in seed drills are explained below:

- **Fluted Feed Type:** The fluted wheel, also referred to as a fluted roller, is powered by a square shaft. It has horizontal grooves along its outer edge, and the wheel can be adjusted sideways to regulate the seed rate. These rollers are positioned at the bottom of the seed box, where they collect seeds in the longitudinal grooves and transfer them to the seed tube through the seed hole.
- **Internal Double Run Type:** The internal double run feed mechanism features a dual-faced wheel. One side has larger openings for coarse seeds, while the other side has smaller openings for fine seeds. A flapper gate at the bottom of the box covers the unused opening. The seeding rate is adjusted by altering the speed of the internal feed wheels, which is accomplished by engaging the appropriate gears.
- **Cup Feed Mechanism:** The mechanism consists of a circular shaft with multiple circular discs mounted on it. Each disc is equipped with several cups or spoons arranged along a circular path. This setup is positioned at the bottom of the seed box. As the shaft rotates, the discs also turns. The cups collect the seeds and then release them into a funnel, from which they travel through the seed tube and fall into the furrows. The cups have two faces: one designed for larger seeds and the other for smaller ones.
- **Cell Feed Mechanism:** This mechanism uses a series of evenly spaced cells engraved on the edge of a circular plate or wheel to collect and deliver seeds.

- **Brush Feed Mechanism:** A rotating brush controls the flow of seeds from the hopper into the seed tube. This mechanism is commonly found in bullock-drawn planters in the country.
- **Picker Wheel Mechanism:** In this mechanism, a vertical plate with radially extending arms picks up seeds and drops them into the furrow. It is typically used for larger seeds such as potatoes.
- **Star Wheel Mechanism:** This feed mechanism features a toothed wheel that rotates horizontally and conveys fertilizer through a feed gate beneath the star wheel.
- **Auger Feed Mechanism:** This mechanism uses an auger to move material from the container and distribute it uniformly in the field. Fertilizer drills often use this mechanism, and some seed drills also incorporate it.

Maintaining Seed Metering

- **Regular Cleaning:** Seed metering mechanisms, especially fluted rollers and precision plates, can get clogged with dirt, debris, or leftover seeds. Regular cleaning ensures smooth operation.
- **Lubrication:** Moving parts, especially gears and rollers, should be properly lubricated to prevent wear and tear.
- **Calibration:** Regular calibration is necessary to maintain accurate seed rate and spacing. This can be done by adjusting the settings or checking seed flow with test runs.

Merits of a Good Seed Metering Mechanism

- **Accurate Seed Placement:** Ensures that seeds are placed uniformly in the soil at the correct depth and spacing.
- **Reduced Seed Wastage:** Proper metering ensures that only the necessary amount of seed is dispensed, minimizing waste and reducing costs.
- **Improved Crop Uniformity:** Even distribution of seeds leads to a uniform crop stand, which improves overall productivity.
- **Increased Germination:** Proper seed depth and spacing enhance seed-to-soil contact, improving germination rates and early-stage growth.

TYPES OF FURROW OPENERS

Furrow openers are essential components in a seed drill, designed to open a furrow in the soil. Seeds travel through the seed tube and are deposited into these furrows for proper planting. Various types of furrow openers are in use, including:

- Shovel type
- Shoe type
- Hoe type
- Disc type (single disc, double disc), and others.

Shovel Type

Shovel-type furrow openers are widely used in seed drills. There are three variations of shovels commonly in use: (a) reversible shovel, (b) single point shovel, and (c) spearhead shovel. These openers are ideal for stony and root-infested fields, providing efficient furrow creation. The shovels are bolted to the shanks at the bottom, and boots are attached to the back of the shovels. These boots carry the delivery ends of the seed tubes, guiding seeds into the furrow. In cultivator-type seed drills, shovel-type furrow openers are commonly employed due to their versatility and durability in challenging soil conditions.

Shoe Type

Shoe-type furrow openers are generally effective in trashy soils or when the seed beds are not smoothly prepared. Constructed from two flat pieces of steel welded together to form a cutting edge, they are especially well-suited for black soils. Bullock-drawn, three-tyne seed drills typically use shoe-type furrow openers. These openers are designed to efficiently penetrate rougher soil conditions and help create a furrow for seed placement.

Disc Type

Disc-type furrow openers come in two main variants:

Single Disc Type

Single disc furrow openers are perfect for fields where plant residues or trash are used as mulch. These openers consist of a curved, hardened steel disc set at an angle. As the disc moves through the soil, it shifts the soil to one side, forming a small furrow. Seeds are then placed into this furrow. To keep the disc clean and free from soil buildup, two scrapers are employed: one toe-shaped scraper on the convex side and one 'T'-shaped scraper on the concave side. These openers work well in sticky soils, although they can be costly and require more maintenance due to the precision and materials involved.

Double Disc Type

Double disc furrow openers feature two flat discs positioned at an angle to one another. These discs work together to create a clean furrow while leaving a small ridge in the center. The seeds are placed between the two discs, providing more accurate and uniform seed placement. This type of opener is particularly useful for trashy lands or fields with heavy plant residues. Seed drills that operate at higher speeds often employ double disc furrow openers due to their efficiency in maintaining clean furrows and seed placement.

Seed Tube

Seed tubes are located at the lower end of the feed cups and are responsible for conducting seeds from the feed cups into the furrow lines. These tubes guide the seeds through suitable boots and furrow openers. Seed tubes are typically made from materIals such as polypropylene, rubber, or steel. The least diameter of both the seed and fertilizer tubes is 25 mm to ensure proper seed flow and distribution.

Boot

A boot is a component of the sowing equipment that receives the seeds or fertilizers from the seed tube and delivers them to the furrow. It is generally constructed from cast iron or mild steel to ensure durability and proper functioning in the field. The boot plays a crucial role in the accurate placement of seeds into the furrow, ensuring efficient sowing operations.

VARIANTS OF FURROW CLOSERS

Drag chain, drag bars, scraper blades, steel press wheels, zero- pressure pneumatic press wheels, disc hillers and various combination of these are used as furrow closers.

Figure 5.3: Furrow Closers

Planter

A planter is a piece of sowing equipment designed for planting seeds that are larger in size and cannot be managed by seed drills. Planters ensure precise row-to-row and plant-to-plant spacing, which is essential for optimal crop growth. Popular types of planters include potato planters, maize planters, and cotton planters.

Functions of a Planter

- To open the furrow
- To meter the seed

- To deposit the seed in the furrow
- To cover the seed and compact the soil over it.

Components of a Planter

A planter is made up of several essential components, including:

- Hopper
- Feed metering device
- Knock-out mechanism
- Cut-off mechanism
- Furrow opener
- Furrow closer
- Drive mechanism
- Clutch, etc.

Each row in a planter is equipped with a separate seed hopper, which is generally constructed from mild steel or another suitable material.

Advantages

- More uniform crop emergence due to even soil coverage over each seed row, reducing the issue of uneven growth at harvest time, such as green streaks in the crop.
- Helps fill in surface trenches, leaving the field smoother, which reduces the need for future equipment repairs.
- Features a K-Hart trip assembly that prevents closer damage when encountering rocks, with up to 10″ of trip travel.

Features & Specifications

- Adjustable angle blade.
- Optional smooth blades.
- Polyurethane bushings in the trip assembly.
- Proven bearing design for years of maintenance-free use, featuring a double-row ball assembly.
- Blades made from special wear-resistant material, ensures durability.
- Weight: 22 lbs.
- Color: Gloss black.

Seed Metering Device in a Planter

The seed metering device in a planter is an important component that ensures the precise placement of seeds at the correct depth and spacing during planting. It regulates the flow of seeds from the seed hopper to the seed tubes, ensuring uniform distribution across the field. Proper operation of the seed metering device is vital for uniform germination and optimal plant growth, contributing to a high crop yield.

Various seed metering devices are used in planters. The most common type features a rotating circular plate with cells positioned at the bottom of the seed hopper. Some planters also use vertical rotors, inclined rotors, or canvas belts. The circular seed plates have notches or holes along their perimeter, known as cells, which pick up seeds from the hopper and drop them into the seed tube. These rotors are classified based on the type of notches provided: (i) Edge drop, (ii) Flat drop, and (iii) Hill drop rotors.

In flat drop rotors, the cells are shallower and only allow one seed at a time. Hill drop rotors have larger cells that can accommodate multiple seeds. The rotating plate receives seeds from the hopper and passes them under a cut-off mechanism, which ensures that only the seeds accommodated in the cells are allowed to pass through.

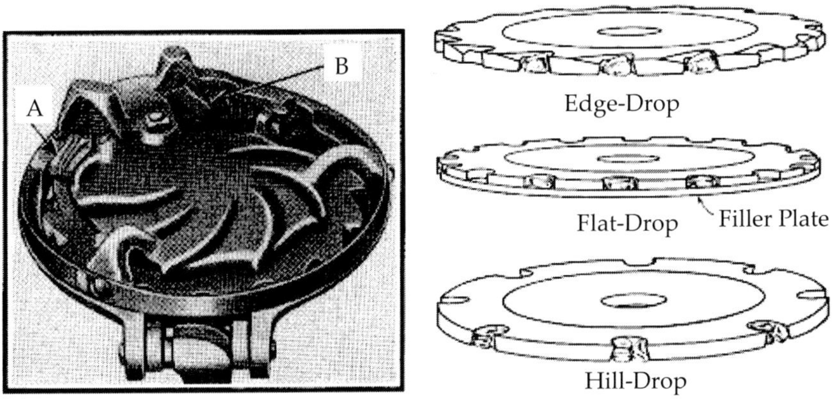

Figure 5.4: Left: Hopper Bottom for Typical Horizontal-Plate Corn Planter. Note the Spring-Loaded (Yielding) Cutoff A, and the Spring-Loaded Knockout Pawl B. Right: Three Types of Edge-Cell Plates used Interchangeably in this Hopper Bottom

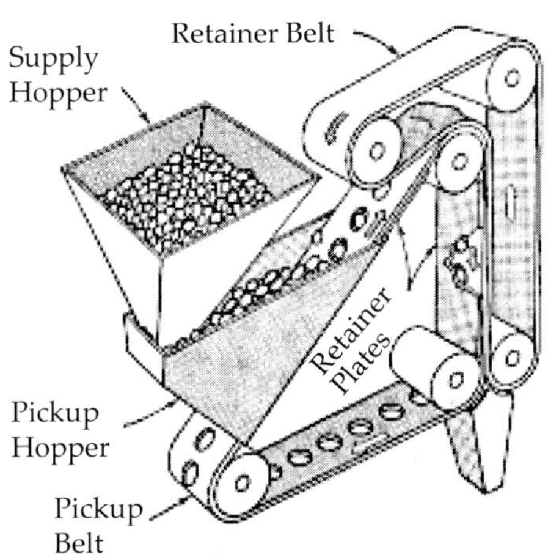

Figure 5.5: Experimental Belt-Type Planter Developed for High-Speed Drilling of Seeds Such as Peanuts

Cut-off Mechanism: The cut-off mechanism serves to eliminate or brush out excess seeds that are carried away by the cells. This ensures that only the correct number of seeds is delivered to the furrow.

Knock-out Mechanism: The knock-out mechanism is responsible for releasing seeds from the rotor cells. It consists of rollers or star wheels that apply pressure to release the seeds, which then fall into the seed tube and are directed to the furrow.

Spacing of Seeds or Hills: The spacing of seeds or hills along the row is determined by the ratio of the peripheral speed of the rotor to the forward speed of the planter, as well as the spacing between the cells on the rotor.

Factors Affecting Planter Accuracy: The accuracy of a planter depends on several factors, including:

(i) Speed of the seed plate,

(ii) Shape and size of the cells,

(iii) Shape of the hopper bottom,

(iv) Uniformity of seed size.

Roles of Seed Metering Device in a Planter

- **Uniform Seed Spacing:** Ensures that seeds are placed at regular intervals, preventing overcrowding or excessive spacing between plants.
- **Correct Seed Depth:** Maintains consistent planting depth for seeds, which is essential for uniform germination.
- **Seed Rate Control:** Regulates the number of seeds being sown per unit area, allowing farmers to adjust the seeding rate based on crop and field conditions.
- **Minimization of Seed Wastage:** A well-calibrated seed metering device ensures that seeds are used efficiently, minimizing waste and reducing costs.
- **Adaptability:** Allows adjustments for different seed sizes, shapes, and types of crops.

POTATO PLANTER

Potatoes are a significant crop in India, and planting them requires considerable labor. There are two main types of potato dropping mechanisms in potato planters:

- Automatic, and
- Semi-automatic.

Potato Planter (Automatic Seed Dropping)

The automatic potato planter consists of a hopper for each row and cups connected to a chain drive mechanism. The graded potatoes are picked up by the cups and carried to the furrow opener spout, where they are released into the furrows. A feeder roller, attached to the compensating tray holding extra potatoes, ensures uniform seed spacing by releasing a potato into an empty cup. The planter can simultaneously apply fertilizer and pesticide. It can plant 2 to 4 rows, with a capacity of 6000 to 14000 potatoes per hour.

Potato Planter (Semi-automatic)

The semi-automatic potato planter consists of a hopper, metering disc, and furrow openers. A separate fertilizer application unit can be attached. Potatoes from the hopper are placed into compartments of the metering disc, which is rotated by a gear drive mechanism. Furrow openers form ridges, and the potatoes drop into the furrows by gravity. This type of planter can plant 2 to 4 rows, with a field capacity of 0.15 to 0.35 hectares per hour.

Calibration of Seed Drill

Calibration of a seed drill involves laboratory testing to determine the rate of seed delivery. This process ensures that the seed drill delivers seeds according to the recommended seed rate. If discrepancies are found between the actual seed delivery and the recommended rate, adjustments are made to the seed metering mechanism to correct the delivery rate. Calibration helps avoid the difficulty of setting the seed delivery rate in field conditions.

TRACTOR IMPLEMENTS

The adoption of tractor implements has revolutionized farming practices, offering numerous benefits, including reduced labor costs and increased production efficiency. Among these innovative tools, the Vacuum Planter stands out as one of the advanced planting technologies provided by John Deere India, showcasing creativity and precision in farming.

The Super Seeder

The Super Seeder exemplifies efficiency by merging three crucial tasks—tilling, sowing, and seedbed covering—into one seamless operation. Specifically designed for wheat sowing in fields after paddy harvesting, this cutting-edge implement not only enhances farm productivity and profitability but also plays a key role in promoting environmental sustainability by reducing the need for burning paddy straw.

Figure 5.6: Super Seeder

Features & Benefits

- Single-pass solution for wheat sowing
- Environmentally friendly and efficient operation
- Double furrow disc opener for precise seed placement

Seed Cum Fertilizer Drill

The Seed Cum Fertilizer Drill is engineered to sow a variety of seeds while simultaneously applying fertilizer, providing exceptional versatility. Whether anyone is planting wheat, corn, oilseeds, or pulses, this implement guarantees optimal seed placement and fertilization across different soil types.

Figure 5.7: Seed Cum Fertilizer Drill

Features & Benefits:

- Seed and fertiliser rate adjustment mechanism
- Adjustable row spacing for customised planting
- Depth adjusting wheel for precise depth control

Multi-Crop Vacuum Planter

The Multi-Crop Vacuum Planter is a benchmark in precision planting, offering exceptional accuracy and minimal seed wastage. It is Capable of sowing a variety of crops, including cotton, corn, soybean, and grains, this advanced equipment ensures optimal yields while keeping maintenance costs low, making it an essential tool for farmers aiming for high efficiency.

Figure 5.8: Multi-Crop Vacuum Planter

Features & Benefits

- High rate of germination
- Low maintenance costs
- Compatibility with all soil types

Roto Seeder

The Roto Seeder seamlessly integrates the functions of a rotary tiller and a seeder, offering a versatile and efficient solution for wheat planting. With its cost-effective design, this equipment ensures uniform seed placement, minimizes waste, and plays a pivotal role in enhancing productivity in modern farming.

Figure 5.9: Roto Seeder

Features & Benefits

- Efficient utilization of time and workforce
- Cost-effectiveness and versatility
- Compatibility with all soil types

Mechanical Planter

The Mechanical Planter is ideal for planting a wide range of crops across various soil types, ensuring uniform seed spacing and depth. Its precision reduces seed wastage and boosts germination rates, making it highly efficient. Additionally, it is fully compatible with most of the tractors in the market, enhancing its versatility and performance.

Figure 5.10: Mechanical Planter

Features & Benefits

- Higher rate of germination
- High precision of seed placement
- Compatibility with multiple crop types

MINIMUM TILLAGE EQUIPMENT

Minimum tillage is a farming practice aimed at reducing soil disturbance compared to conventional tillage. It involves using specialized equipment designed to perform the necessary soil preparation for planting with minimal disruption to the soil structure. This type of tillage helps to conserve moisture, reduce soil erosion, improve organic matter retention, and promote beneficial soil organisms.

Below are the key constructional details, adjustment procedures, and working principles of minimum tillage equipment.

Structure and Design of Minimum Tillage Equipment

The design and components of minimum tillage equipment are optimized to reduce soil disturbance while ensuring sufficient seedbed preparation. Some common equipment used for minimum tillage includes chisel plows, disk harrows, cultivators, and no-till planters.

Chisel Plow (Subsoiler)

- **Frame:** Made of heavy-duty steel, designed to support the load of the tines and shanks. Often equipped with a flexible frame to allow adjustment to different soil types.
- **Tines/Shanks:** Heavy-duty, sharp, and pointed, the shanks are spaced to penetrate deep into the soil to break compacted layers, but without inverting the soil.
- **Points:** Replaceable tips at the end of each shank, designed for optimal penetration and minimal soil disturbance.
- **Depth Control:** Adjustable wheels or coulters at the rear of the unit control the depth of penetration.
- **Attachments:** Optional attachments like rolling baskets or harrow bars may be used for surface smoothing and residue management.

Disk Harrow

- **Frame:** Typically made from high-strength steel with adjustable widths.
- **Disks:** Large, concave metal disks that break the soil surface. These can be either solid or notched, depending on the intended use. The disks are arranged in gangs at varying angles to till the soil in a controlled manner.
- **Depth Control:** Depth is controlled via adjusting the gang angles or using depth wheels.
- **Gang Assembly:** Can be adjusted in terms of angle and spacing to achieve the desired tillage depth and residue management.

No-Till Planters

- **Frame:** Similar to a traditional planter but designed to minimize soil disturbance. Strong and sturdy, often built with heavy-duty components to manage residue and large amounts of soil pressure.
- **Opening Discs:** These narrow, sharp discs open a slit in the soil for seed placement. They are designed to minimize soil disturbance.
- **Seed Distributors:** These ensure accurate seed placement at the correct depth.
- **Press Wheels:** After the seed is placed, press wheels cover the seed, providing slight compaction to ensure good seed-to-soil contact.
- **Depth Control:** Adjustable components that ensure proper seed depth under varying soil conditions.

Adjustment of Minimum Tillage Equipment

Adjusting minimum tillage equipment is crucial for optimizing performance across different soil types and conditions. The main adjustments include:

Depth Adjustment

- **Chisel Plows:** The depth of penetration can be adjusted using depth wheels or by varying the angle of the shanks. Typically, the depth should be between 6 to 10 inches, depending on soil compaction.
- **Disk Harrows:** Depth control is adjusted by changing the angle of the disk gangs or adjusting the pressure from the tractor's hydraulic system.
- **No-Till Planters:** Seed depth is adjusted using depth control wheels or a mechanical control mechanism to ensure that the seed is placed at the correct depth for optimal germination.

Tine or Disk Angle

- **Chisel Plows:** The angle of the tines (or shanks) can be adjusted to control the level of soil disturbance. The more vertical the shanks, the less disturbance to the soil surface.
- **Disk Harrows:** The angle of the disk gangs controls the level of tillage intensity. A more aggressive angle increases soil disturbance, while a shallow angle is less disruptive.

Residue Management

- **Chisel Plows:** Optional residue managers (e.g., rolling baskets) can be attached to the rear to break up large clumps of residue, which improves surface leveling.
- **Disk Harrows:** Can be adjusted to manage residue, with varying disk angles helping to either bury or leave residue on the surface.
- **No-Till Planters:** May include residue management features such as row cleaners or coulters to move residue aside and allow proper seed placement.

Operation of Minimum Tillage Equipment

The operation of minimum tillage equipment focuses on minimal disturbance of the soil, ensuring that only the necessary tasks are completed for seed planting and soil health. Below is an overview of how each equipment works:

Chisel Plow

- **Working Principle:** The chisel plow works by breaking up compacted soil layers beneath the surface without turning over the soil. The shanks penetrate the ground at an angle, fracturing the compacted layers. This allows water and air to penetrate deeper into the soil while leaving surface residues intact.
- **Benefits:** This practice encourages root growth, improves water infiltration, and minimizes erosion by leaving organic matter on the soil surface.

Disk Harrow

- **Working Principle:** The disk harrow works by slicing through the soil with its rotating disks. These disks break the soil surface and incorporate crop residues into the soil. The depth and angle of the disks determine the level of soil disturbance.
- **Benefits:** Ideal for shallow tillage to prepare a seedbed without disrupting deeper soil layers. The disks can also be used to manage surface residue and help with seedbed leveling.

No-Till Planter

- **Working Principle:** No-till planting equipment is designed to plant directly into undisturbed soil, cutting a small slit for the seed to be placed. The openers (discs or coulters) cut through surface residue and soil, creating a small furrow. The seed is dropped into the furrow, and press wheels close it, ensuring good seed-to-soil contact.
- **Benefits:** Minimal soil disturbance reduces erosion, improves moisture retention, and enhances soil organic matter. It also helps in maintaining soil structure.

Advantages of Minimum Tillage Equipment

- **Reduced Soil Erosion:** By disturbing the soil less, erosion is minimized, especially on sloped fields.
- **Increased Organic Matter:** Residue left on the soil surface degrades slowly, adding organic matter to the soil and improving its fertility.
- **Improved Soil Structure:** No-till and minimum tillage help preserve soil aggregates, enhancing water infiltration and root penetration.
- **Lower Fuel Consumption:** Less intensive tillage requires less energy and fewer passes over the field, reducing fuel consumption.
- **Water Conservation:** Less evaporation and better retention of soil moisture.

Final Thoughts

Minimum tillage equipment plays a crucial role in modern agricultural practices that aim to conserve soil and improve long-term farm sustainability. Proper adjustment and understanding of how each piece of equipment works are crucial for optimizing crop production and maintaining soil health. As farming practices continue to evolve, the role of minimum tillage equipment becomes increasingly important in addressing environmental concerns and achieving high-efficiency, low-impact agriculture.

6

Adjustments and Calibration of Seed Drills

CALIBRATION OF SEED DRILL

Calibration of a seed drill refers to the process of testing the seed drill to ensure it delivers the correct seed rate. Before using the seed drill in the field, calibration is essential to achieve the desired seed rate. This procedure is done to ensure that the machine delivers the predetermined seed rate accurately. The following steps are followed for the calibration of a seed drill.

Procedure for Calibration of a Seed Drill

- Determine the nominal width (W) of seed drill .

$$W = M \times S,$$

Where, M = Number of furrow openers

S = Spacing between the openers, m

- Find the length of the strip (L) having nominal width (W) necessary to cover 1/25 ha (1 /25 x 10000 m^2).

$$L = 400/W \text{ meter}$$

- Determine the number of revolutions (N) of the ground wheel of the seed drill required to cover the length of the strip (L).

$$L = P \times D \times N = 400/W$$

$$N = 400/P \times D \times W \text{ revolutions per minute}$$

- Jack the seed drill so that the ground wheels turn freely. Make a mark on the drive wheel and a corresponding mark at a convenient place on the body of the drill to help in counting the revolutions of the ground wheel.
- Fill the selected seed in the seed hopper. Place a container under each boot for collecting the seeds dropped from the hopper.

- Set the seed rate control adjustment for maximum position and mark this position on the control for reference.
- Engage the clutch and rotate the ground wheel for N = 400/P x D x N, revolutions per minute.
- Weigh the quantity of seed collected in the container and record the observation.
- Calculate the seed rate in kg/ha.
- If the calculated seed rate is higher or lower than the desired rate of selected crop, repeat the process by adjusting the seed rate control adjustment till the desired seed rate is obtained.

Key Objectives of Drill Calibration

Determining the Area Seeded During Calibration Distance

To calibrate the seed drill, measure a 200-foot distance in the field. This distance is selected because it's practical yet long enough to observe field variability. For example, if your drill has a working width of 10 feet, the area seeded would be 2,000 square feet (200 feet by 10 feet). To find the fraction of an acre, divide the 2,000 square feet by 43,560 square feet per acre, resulting in an area of 0.046 acres.

Determining the Number of Revolutions of the Drive Wheel

In the field, measure a 200-foot distance using a tape measure or other device. Identify the wheel responsible for driving the seeding mechanism. This could be a special wheel or one of the support wheels. Count how many times this wheel rotates while covering the 200-foot distance. Mark the wheel to make it easier to track the number of revolutions. Repeat this process for accuracy and average the results.

Setting Up the Drill for Calibration

Stationary calibration simulates the drill's operation by turning the drive wheel a set number of times. Raise the drive wheel above the ground using a hydraulic jack so that it can spin freely. Secure the drill to the tractor for safety. Use the manufacturer's rate chart to adjust the seed meter to the desired seed rate. If the specific seed is not listed, select a similar seed size. Ensure the seed cups are full, and the seed metering mechanism is covered. Use a tarp or containers to collect the seed from the openers, ensuring no obstruction in the seed tubes. Turn the drive wheel the necessary number of times to simulate 200 feet of travel. Collect the seed from the tarp or containers and weigh the total. Use a gram scale and convert the weight to pounds by dividing by 454 grams per pound. If the collected seed represents a lower or higher seeding rate than desired, adjust the metering mechanism and repeat the process. The total seed weight divided by the area drilled will give the seeding rate. For example, 2.3 pounds of seed over 0.046 acres equals a seeding rate of 50 pounds per acre. Check individual openers for consistency. If any opener deviates by more than 10 percent from the average, it may require maintenance.

Figure 6.1: Drill Calibration

Factors Affecting Seeding Rate and Seed Purchase

When determining seeding rates, several factors need to be considered. A reliable source for seeding rate recommendations is the state's agricultural Extension office. Seed suppliers often provide specific seeding rates for their species and varieties, which may be adjusted based on factors such as planting date, soil type, whether planting is done in tilled or no-tilled conditions, soil fertility levels, and other environmental factors. Typically, recommended seeding rates are designed to achieve a desired plant population after emergence.

Seeding rates in pounds per acre are calculated based on the number of seeds in one pound of seed. Once this information is known, adjustments should be made for percent germination and the amount of non-seed material indicated on the seed label. Hard seeds, that are seeds that germinates more slowly than pure seeds, are common in legumes. However, seed rates typically do not require modification for hard seed unless it exceeds 5 percent. In such cases, consulting the seed supplier is recommended.

WORKING MECHANISM OF SEED DRILLS

The seed metering mechanism of a seed drill or fertilizer distributor is responsible for delivering seeds or fertilizers from the hopper at selected rates. Several types of seed metering mechanisms are used, including fluted roller feed, internal double run type, cup feed, cell feed, brush feed, auger feed, picker wheel, and star wheel mechanisms.

- **Fluted Roller Feed Type:** This mechanism uses an adjustable fluted roller to collect and deliver seeds into the seed tube. The fluted roller, mounted on a shaft, rotates and carries seeds into longitudinal grooves. These grooves pass the seeds through holes into the seed tube. By adjusting the position of the fluted wheel, the length of the exposed grooves can be modified, thus controlling the amount of seed delivered. This mechanism provides more precise metering compared to the internal double run type.

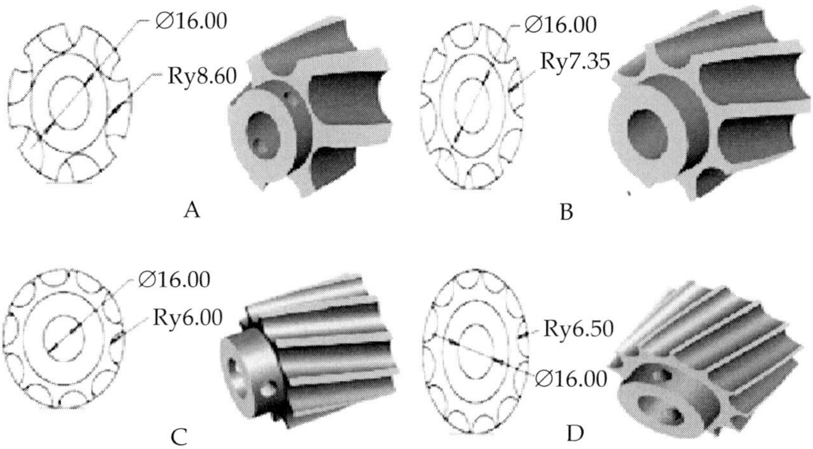

Figure 6.2: Fluted Roller Type

- **Internal Double Run Type:** The internal double-run mechanism features a feed wheel with fine and coarse ribbed flanges. The wheel has two faces, one with larger openings for large seeds and the other with smaller openings for small seeds. A gate at the bottom of the seed box covers the unused openings. The seed rate is adjusted by changing the speed of the spindle that carries the discs.

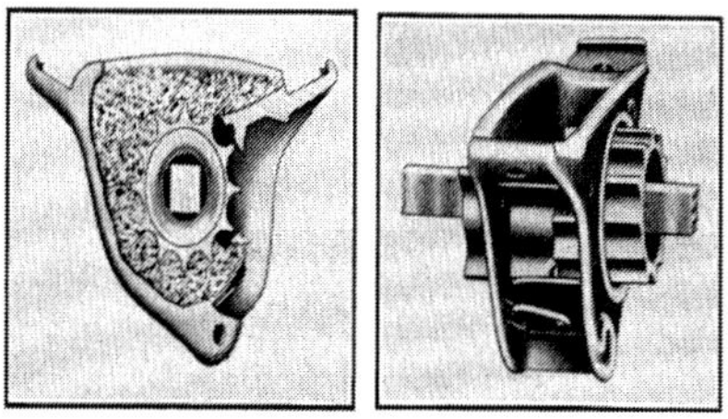

Figure 6.3: Internal Double Run Type

- **Cup Feed Mechanism:** This mechanism consists of cups or spoons attached to a rotating vertical disc, which picks up seeds from the hopper and delivers them into the seed tubes. The seed hopper has two parts: the grain box (upper) and the feed box (lower). The rotating spindle carries discs with cups, which pick up seeds and drop them into small hoppers. The rate of seed delivery is controlled by the speed at which the seed barrel revolves, and this mechanism is common in British seed drills.

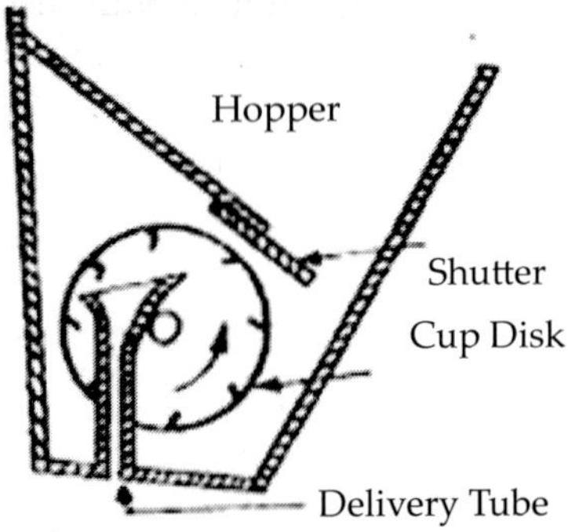

Figure 6.4: Cup Feed Mechanism

- **Cell Feed Mechanism:** Seeds are collected and delivered by a series of evenly spaced cells located on the periphery of a rotating circular plate or wheel. Each cell picks up a certain amount of seed and deposits it into the seed tube.
- **Brush Feed Mechanism:** In this mechanism, a rotating brush regulates the flow of seed from the hopper. It is commonly used in bullock-drawn planters in some regions.
- **Auger Feed Mechanism:** The auger feed mechanism uses an auger to distribute seeds or fertilizers evenly. The auger helps in moving the material through an aperture at the base or side of the hopper. It is commonly used in fertilizer drills.
- **Picker Wheel Mechanism:** The picker wheel mechanism features a vertical plate with radially projected arms that help drop large seeds, like potatoes, into furrows. Suitable jaws are used for accurate seed placement.
- **Star Wheel Mechanism:** The star wheel mechanism uses a toothed wheel that rotates horizontally. It conveys fertilizer through a feed gate located below the star wheel. This system is used for precise fertilizer distribution.

ADJUSTMENT AND MAINTENANCE OF SEEDING AND PLANTING EQUIPMENT

Seeding Equipment/Implement

Seeding equipment refers to machines used for sowing seeds at the required row-to-row spacing and depth. These machines are designed to ensure precise seed placement in the soil, facilitating optimal germination and growth conditions. Seeding implements typically include various mechanisms for seed distribution, such as furrow openers, seed tubes, and metering devices. The primary function of seeding equipment is to plant seeds at the correct depth and spacing, ensuring uniformity and maximizing crop yield.

Planting Equipment/Implement

Planting equipment is a machine used for planting seeds at the required row-to-row spacing, plant-to-plant spacing, and depth. These implements are more specific than seeding equipment, as they cater to both seed and plant placement. Planting machines are designed to accommodate different types of seeds or seedlings, adjusting the spacing and depth to meet specific crop requirements. They often include features like adjustable seeders, plant spacing controls, and depth regulators to ensure consistent planting.

Figure 6.5: View of Zero Till Multi-Crop Planter

Adjustments in Multi-crop Planter

Furrower openers and ridgers

Furrow openers can be adjusted by removing or raising a few furrow openers with shanks according to the number of rows and spacing in different crops to be raised on the top of the raised beds. Three ridgers have been provided to make two beds in a single pass. Most farmers have tractors in the range of 35-45 HP, and these tractors can easily pull the three ridgers. The height of the beds in wheat varies from 15 to 30 cm and can be further adjusted. The width of the furrow and top of the bed generally remains 30 cm and 37 cm, respectively, though it can be varied from 20 to 40 cm by adjusting the wing width.

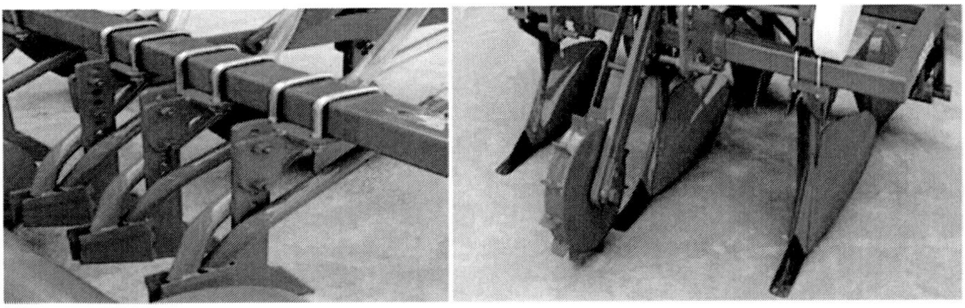

Figure 6.6: View of Ridger Spacing or View of Ridgers

Figure 6.7: View of Backside of Ridger

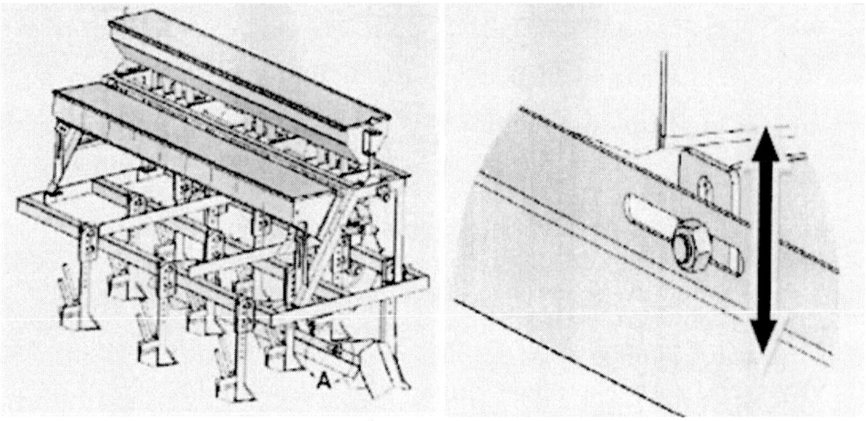

Figure 6.8: View of Adjustment of Bed Shaper

Adjustment of the Bed shaper

The Bed Shaper should also be adjusted in the upward and downward positions for shaping the bed. To adjust, loosen the bolt and lift the shaper to the upward position if a higher bed is required. Then, tighten the nut and bolt again. For a lower bed position, loosen the nut and bolt once more, shift the shaper downward, and tighten the nuts and bolts.

Seed Metering Speed Adjustments

As shown in Figures, select appropriate combination of gear/sprockets for getting desired speed of rotation of seed &fertilizer metering shafts.

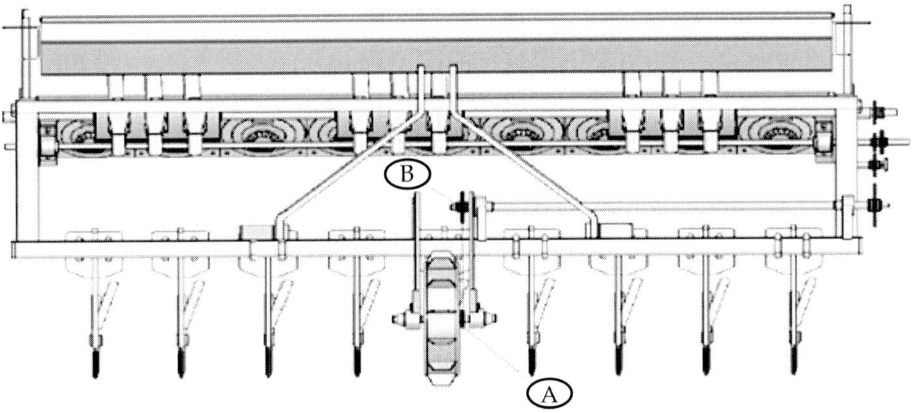

Figure 6.9: View of Backside of Planter

The below figure shows following processes:

A	11 Teeth sprocket
B	20 Teeth sprocket
C	28 teeth sprocket
D	11Teeth sprocket for seed box
E	15 Teeth sprocket for fertilizer box
F	15 Teeth idle
G	Transmission chain

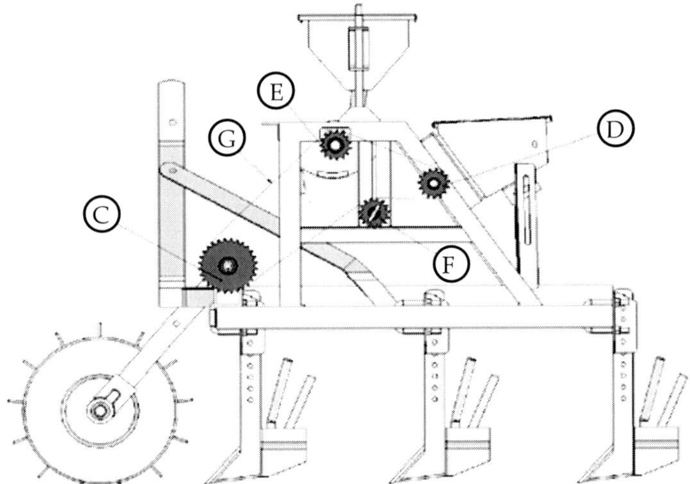

Figure 6.10: Side View of Planter

Seed Rate Adjustments

- In order to adjust the inclined plates we have to put washer between the nylon plate and housing on the shaft so that the face of plate should not have any gap between the face of housing.
- The function of press plate is to provide proper pressure on the ring.
- Adjust the nut as per required pressure on the ring.

Increasing and Decreasing the Flow of Seed

To increase the flow of seed, lift the seed box in the upward direction, and to decrease the flow shift the seed box in the downward direction .

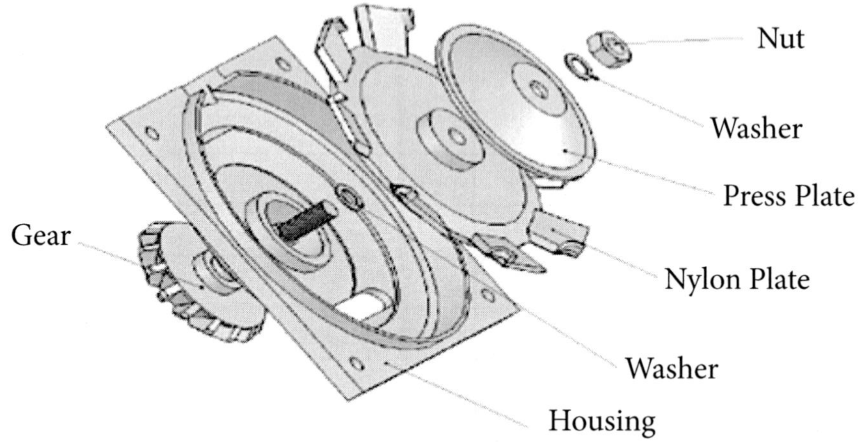

Figure 6.11: View of Assembly of Nylon Plate and Housing

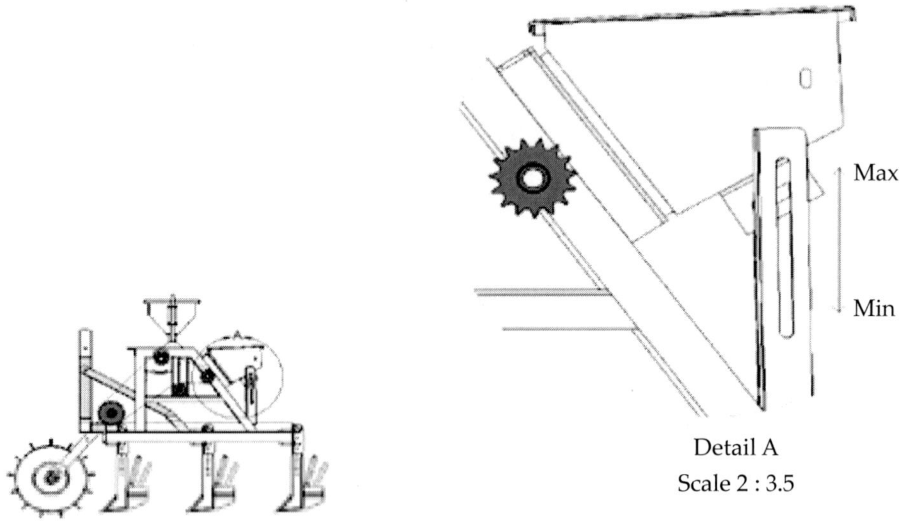

Figure 6.12: View of Seed Increase and Decrease on Seed in Plater

Seed Metering

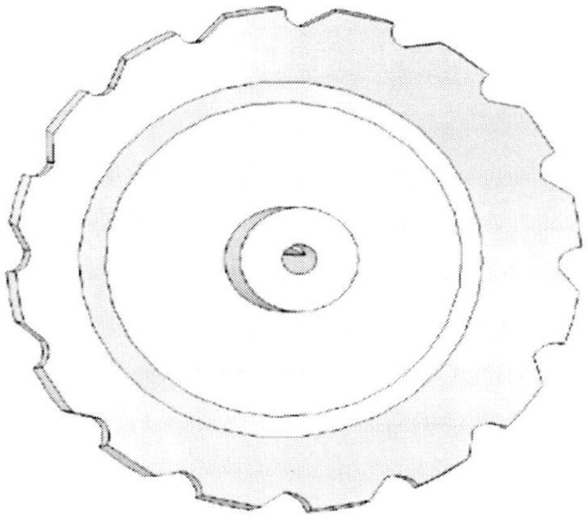

Figure 6.13: View of 16 Teeth Sprocker for Wheat

For Wheat

- Use the 16 no`s spoon plane plate.
- As given in the Figure on page 123 the sprocket section is as under:

Position	
C	28 Teeth sprocket
D	11 Teeth sprocket for seed box
E	15 teeth sprocket for fertilizer box
F	15 teeth idle

For Soyabean

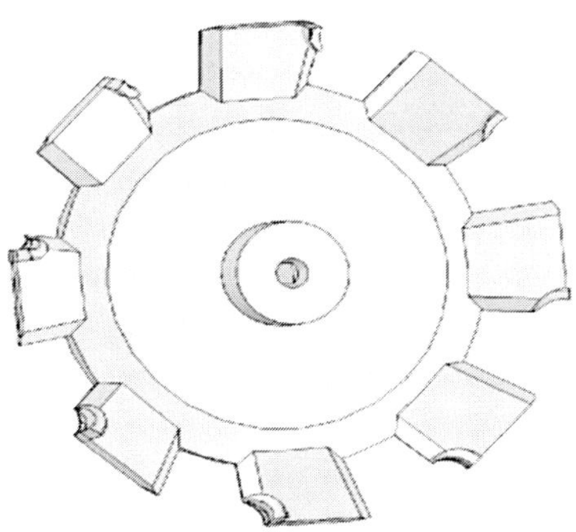

Figure 6.14: View of Sprocket for Soyabean

- Use the 8 no`s spoon inclined plate no. "0".
- As given in above figure sprocket selection is as under:

Position	
C	28 Teeth sprocket
D	15 Teeth sprocket for seed box
E	15 Teeth sprocket for fertilizer box
F	15 Teeth idle

For Maize, Chick Pea and Bengal Gram

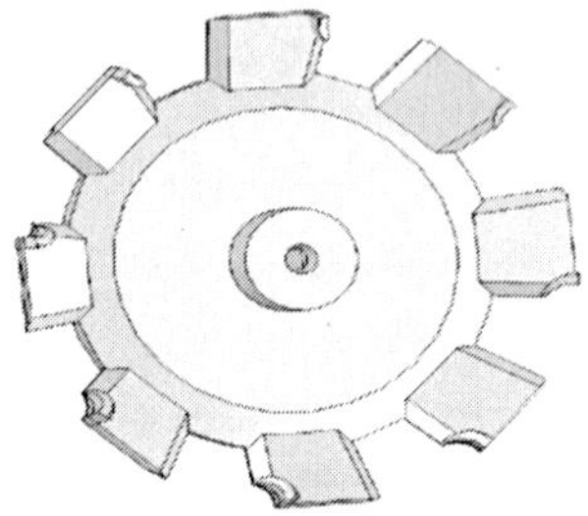

Figure 6.15: View of Sprocket for Maize, Chick Pea and Bengal Gram

- Use the 8 no`s spoon inclined plate no. "20".
- As given in below figure sprocket selection is as under:

Position	
C	11 Teeth sprocket
D	15 Teeth sprocket for seed box

For Ground Nut, Hybrid Maize

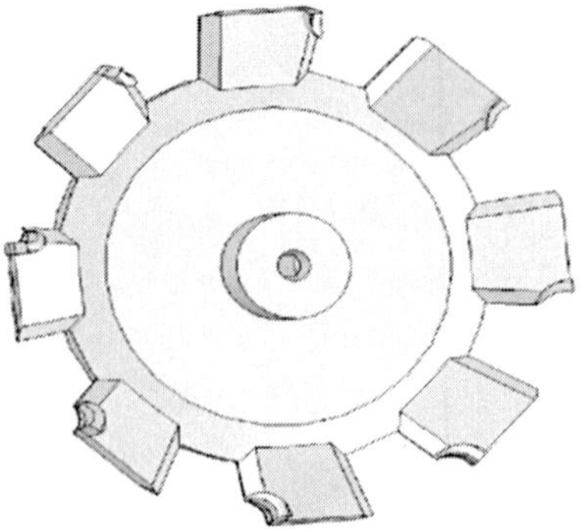

Figure 6.16: View of Sprocket for Ground Nut, Hybrid Maize

- Use the 8 spoon inclined plate no. "30".
- Sprocket selection is as under:

Position	
C	15 Teeth sprocket
D	28 Teeth sprocket for seed box
E	15 Teeth sprocket for fertilizer box
F	15 Teeth idle

Calibration of Seeding Equipment

Before taking the seed drill or crop planter into the field it is needed to calibrate its metering mechanism to get the required seed rate per hectare or per acre of the crops which is to be planted. The following steps are followed in calibration of the seed drill:

1.Jack up the seed drill and mark a reference point on the ground wheel.

2.Measure the diameter of ground wheel. Let us say it is "D"meter.

3.Measure the working width of the drill or planter and let us say it is 'W' meter.

$$W = n \times d$$

Where, n = no. of tynes

d = distance between two tynes (m)

4.Workout the circumference of ground wheel (x D)

5.We assume that the planter or drill has to be used in a field having length x breadth as 100 x 100 sq. m.

6.Workout the revolution of the ground wheel required to travel a distance of 100 m.

$$= 100/(\pi \times D)$$

$$= X \text{ (say)}$$

7.Now give the 'X' revolution to the wheel and collect the seeds from all the furrow openers separately. Weigh all the collected seed separately. It should be noted that there should not be large variation in the weight of the seed collected fromthe different furrow openers. Let's say the total weight of the seeds collected from all furrow openers is 'P' kg

8.Workout the total no. of revolutions required to cover one hectare of field:

$$= \frac{100 \times 100}{\pi \times D \times W} = Y \text{ (say)}$$

9.Allow 10% slippage of groundwheel under actual field conditions and recalculate 'Y'.

10.Workout the total amount of seed for 'Y' revolutions

For 'X' revolutions 'P' kg of seed are collected

Therefore, for Y revolution = ------------------- = G (kg)

X x Y

Thus, by substituting the value of 'P' and 'X' revolutions we can obtain the quantity of seed to be sown in the field. Thus, the process is repeated by suitably adjusting the lever on the indexing device till we get the desired seed rate, recommended for a particular crop under planting or sowing.

Planting Operations

Seed should be of good quality and free from dirt and dust.

- Fertilizer should not have clods; clods must be properly broken to a uniform size for free flow.
- All nuts, bolts, and springs should be checked thoroughly. Defective parts should be replaced, and nuts/bolts properly tightened.
- Seed and fertilizer boxes should be cleaned thoroughly.
- The multi-crop rings shaft should move freely; otherwise, the plates may damage the seed.
- Seed plates should be thoroughly cleaned, and any blockages must be removed.
- Ensure that plastic pipes do not have excessive bends, as this will block the free flow of seed and fertilizer in tubes.
- The chain sprocket of the metering mechanism should be properly aligned. Appropriate tension should be maintained for the free movement of seed and fertilizer metering shafts. If any noise occurs during operation, stop the machine and check it.
- Furrow openers should be fitted according to the required row-to-row distance for the crop. There should be no crossing or twisting of furrow openers.
- Fill the seed and fertilizer boxes and calibrate the machine. Ensure that the seed drill is set to the desired seed and fertilizer rates to ensure proper metering, resulting in excellent germination, a good crop stand, and higher yield.

Precautions for the Use of Inclined Plate Planter

The field should be leveled and well prepared before operation.

- The upper soil layer should not be allowed to dry up before sowing, as this will require deeper seed placement and affect germination.
- Mechanical weeding or interculture with this machine is possible after making small adjustments to the positions of its tines.
- Special care should be taken with the depth of seeding to avoid germination problems.

- Attack of termites may be a concern in sandy areas; special attention should be given to this issue.
- Ensure the machine is balanced before starting sowing.
- All inclined plate seed boxes should be in a straight position.
- The depth of sowing should be adjusted using the depth control wheel.
- The seed rate can be increased or decreased by changing the sprocket sets.

Safety Precautions for Using a Raised Bed Planter

- The field should be leveled and well prepared before making beds.
- Beds should be made well in advance, and the field should be irrigated to encourage germination of weeds. The weeds can be controlled either mechanically or with the spray of a non-selective herbicide, such as glyphosate.
- Do not allow the upper soil layer to dry up before sowing, as this will require deeper seed placement and affect germination.
- Mechanical weeding or interculture with this machine is possible after minor adjustments to the tines if the crop is shown in two rows/bed.
- Special care should be taken with the depth of seeding to avoid germination issues.
- Attack of termites may also be a concern in sandy areas; special attention should be paid to this.

If there is an imbalance in the machine attached to the tractor or if the ridger's wings are improperly shaped, a soil layer may form on the alternate beds, hampering seed germination. Farmers should ensure the machine is balanced or modify the wings by cutting the outer edges to remove this layer.

UPKEEP AND REPAIR OF SEEDING MACHINES

A well maintained and properly adjusted seeding machine gives trouble free service for a long time. It also helps in timely completion of operations. The following important points may be kept in mind for the maintenance and repair of various components of the seeding machines.

Machine Maintenance

- Lift the machine off the ground so that the drive wheels are able to move without obstruction.
- Empty the seed and fertilizer from the boxes.
- Open the flow gates on both the seed and fertilizer cups.
- Rotate the drive wheel until the seed and fertilizer from all cups have been emptied. Clean the boxes and cups using a cloth or brush.
- Clean the machine rollers, seed, and fertilizer boxes with diesel to prevent rust formation.
- Apply lubricating oil at the appropriate points, including the bushes and the sides of the metering rollers.

Figure 6.17: View of Seed & Fertillizer box From top Open Position.

Drive/Power Transmission System

- Ensure the drive wheel moves freely. If it's stuck, apply grease or oil to the wheel's bushes. Replace the axle if it is bent or worn.
- The drive wheel should be round; repair it if it is bent.
- Make sure the sprockets on the drive wheel and feed shafts (for seed and fertilizer boxes) are properly aligned.
- Ensure that all sprockets are tightened securely on their shafts to prevent them from moving freely on the shafts.
- Feed shafts should rotate smoothly. If they are stuck due to rust, clean them and apply lubricating oil or grease to the bushes.
- Repair or replace any bent drive shafts.
- Thoroughly clean the seed and fertilizer boxes to ensure smooth movement of the feed shafts.
- Tighten the chain and idler sprockets properly to maintain correct chain tension and ensure the mechanism moves freely.
- Replace worn-out parts, loose or broken components, and any worn bushes.

Seed Metering Mechanism

Typically, an inclined plate type seed metering mechanism is used in this seeding machine. The following steps should be taken to repair and maintain it:

- Remove the seed from the small seed boxes by opening the lower flow gates.
- Unscrew and remove the nuts/bolts of the ring from the base plate.
- Remove the ring from the shaft.
- Inspect the rings and spoons attached to the rings.
- When reinstalling the rings, ensure that all rings are evenly spaced within the seed cups. If the spacing is uneven, adjust it by loosening the bolt on the base plate bush to achieve uniform spacing.

- The entire system should move freely. Rotate the sprocket until the desired seed rate is achieved from all the rings.

Fertilizer Metering Mechanism

In the fertilizer metering mechanism, fertilizer can settle on the parts due to environmental moisture, leading to obstructions that hinder the uniform flow of fertilizer. Large particles might also cause blockages. This system requires special attention:

- After seeding a crop, remove the fertilizer from the box and clean the entire system using a brush or cloth.
- If the system is jammed due to corrosion or rust, remove and clean the cell of the fertilizer feeder.
- Ensure that all plastic tubes/pipes are open and unobstructed.
- The levers on both sides of the fertilizer box should move easily to allow proper metering of fertilizer.
- Tighten all the nuts and bolts of the mechanism.

Seed and Fertilizer Tubes

These tubes, typically made of plastic, connect the seed and fertilizer cups to the seed boots. The following precautions should be taken:

- Secure the tubes to the seed and fertilizer cups using clamps to prevent them from coming loose during field operations.
- Protect the tubes from bending or breaking.
- Replace any old or bent tubes.
- Avoid excessive bending of the tubes, as this can obstruct the free flow of seed or fertilizer, leading to uneven application in the field.

Furrow Openers

Furrow openers are attached to the main frame with U-bolts. These components tend to wear out or twist quickly, so they should be regularly checked and repaired. Any worn-out openers should be removed and replaced when necessary.

General Maintenance

- All components of the machine must be painted to protect against rust and wear.
- The machine should be stored in a dry place, protected from rain, dirt, and dust.
- Moving parts should be greased or oiled on a regular basis to ensure the machine operates smoothly and provides long-term service.
- Providing user training can significantly improve the performance and efficiency of the machines.

7

The Rise of Precision Farming Equipment

Precision farming has revolutionized the agricultural sector by enabling farmers to optimize their operations, increase productivity, and reduce environmental impact. Central to this transformation is the incorporation of GPS technology, which has ushered in a new era of smart farming techniques.

Some key advantages of precision farming equipment include:

- Enhanced accuracy in planting and harvesting.
- Reduced input costs by applying fertilizers and pesticides only where necessary.
- Improved crop yield and quality.
- Minimized environmental impact.
- Efficient resource management.

As we delve deeper into the world of precision farming, let's explore the key equipment and technologies reshaping the agricultural landscape.

GPS TECHNOLOGY: THE BACKBONE OF PRECISION AGRICULTURE

GPS technology has become an essential tool in modern farming. It serves as the foundation for numerous precision farming practices, enabling accurate positioning and navigation during various farm operations.

Below mentioned are some ways GPS technology is transforming agriculture:

- **Auto-steering systems:** Allow tractors and other farming vehicles to navigate fields with centimeter-level accuracy, minimizing overlap and enhancing efficiency.
- **Variable rate application:** GPS-guided systems adjust the application of seeds, fertilizers, and pesticides based on precise field mapping and soil data.

- **Yield mapping:** GPS-enabled combines can generate detailed yield maps, helping farmers identify high and low productivity zones within their fields.
- **Field boundary mapping:** Accurate field boundaries can be mapped using GPS, assisting with land management and record-keeping.

Integrating GPS technology into farming operations can substantially improve efficiency and productivity. Many of the latest tractors and implements come with built-in GPS capabilities, making it easier than ever to adopt this technology.

OVERVIEW OF INNOVATIVE FARM IRRIGATION SYSTEMS

Water management plays a crucial role in sustainable agriculture, and modern farm irrigation systems are at the forefront of efforts to conserve water. These systems use precision farming techniques to deliver water precisely where and when it's needed, reducing waste and promoting crop health.

Precision farming techniques can reduce water usage in irrigation systems by up to 30% compared to traditional methods.

Some advanced irrigation technologies include:

- **Drip irrigation:** Directly delivers water to plant roots, reducing evaporation and runoff.
- **Smart sprinklers:** Utilize weather data and soil moisture sensors to optimize watering schedules.
- **Subsurface drip irrigation:** Applies water beneath the soil surface, minimizing evaporation.
- **Precision mobile drip irrigation:** Combines the benefits of drip irrigation with the flexibility of center pivot systems.

Investing in efficient irrigation systems not only conserves water but also supports healthier crops and lowers energy costs. When considering agricultural machinery for sale, look for irrigation equipment that integrates these cutting-edge technologies.

Use of Drones and Remote Sensing in Agriculture

Drones have become an invaluable tool in precision agriculture, offering farmers an aerial view of their fields and providing critical data for informed decision-making. These unmanned aerial vehicles (UAVs) equipped with sensors can perform a variety of tasks:

- Crop health monitoring
- Pest and disease detection
- Soil analysis
- Irrigation assessment
- Yield estimation

By combining drone technology with other precision farming equipment, farmers can gain unparalleled insights into their operations and make data-driven decisions to enhance crop management.

Figure 7.1: Drones and Remote Sensing in Agriculture

SILAGE AND FORAGE MANAGEMENT: ENHANCING FEED PRODUCTION

Efficient silage and forage management is a key component for livestock farmers and those involved in animal feed production. The integration of modern precision farming equipment has significantly transformed this aspect of agriculture, offering innovative solutions that improve harvesting, storage, and feed quality.

Key technologies in silage and forage management include:

- **Precision forage harvesters:** These advanced machines use GPS guidance and yield monitoring to optimize cutting and chopping operations, ensuring better harvest efficiency.
- **Near-infrared (NIR) spectroscopy:** This technology enables real-time analysis of forage quality while harvesting, allowing farmers to make immediate adjustments to improve feed quality.
- **Silage monitoring systems:** Equipped with sensors, these systems track temperature and moisture levels in silage pits, ensuring that the fermentation process remains optimal.
- **Automated baling systems:** These systems enhance efficiency and consistency in bale production, reducing labor costs and improving the overall quality of the forage.

By investing in these advanced technologies, farmers can improve the quality of their silage and forage, which directly translates into better animal nutrition and increased profitability.

CROP STORAGE SOLUTIONS: ENSURING QUALITY AND MINIMIZING LOSSES

Proper crop storage is essential for maintaining product quality and reducing post-harvest losses. With modern crop storage technologies, farmers can create ideal conditions for long-term storage, ensuring that their produce remains fresh and valuable for longer periods.

Some innovative crop storage technologies include:

- **Automated grain monitoring systems:** These systems utilize sensors to track temperature and moisture levels in grain bins, helping to prevent spoilage and preserve the quality of stored grain.
- **Controlled atmosphere storage:** This technology regulates the levels of oxygen, carbon dioxide, and nitrogen to extend the shelf life of fruits, vegetables, and other perishable crops.
- **Smart ventilation systems:** These systems automatically adjust airflow based on external weather conditions and crop requirements, optimizing the storage environment and preventing deterioration.
- **Robotic grain sampling:** By automating the sampling process, robotic systems ensure accurate quality assessments without the need for manual intervention, streamlining the monitoring process.

When considering agricultural machinery for sale, it is essential not to overlook the importance of modern crop storage equipment. Incorporating these technologies can greatly improve storage efficiency, reduce waste, and preserve the integrity of harvested crops.

SUSTAINABLE FARMING PRACTICES: NAVIGATING THE FUTURE OF AGRICULTURE

Sustainable farming practices are increasingly critical as farmers face challenges such as climate change, resource scarcity, and the need to produce more food with fewer inputs. Precision farming equipment plays a crucial role in implementing sustainable farming practices by minimizing environmental impact while maintaining high productivity levels.

Some sustainable farming practices enabled by modern technology include:

- **Conservation tillage:** This practice reduces soil disturbance and prevents erosion, promoting soil health and water retention.
- **Precision nutrient management:** By applying fertilizers only where and when they are needed, this technique reduces nutrient runoff and enhances soil health.
- **Integrated pest management:** This method uses data collected from sensors and drones to apply pesticides more precisely, reducing overall pesticide usage and minimizing environmental harm.

- **Cover cropping:** Growing cover crops between main crop seasons helps improve soil health, reduce erosion, and increase organic matter in the soil.

By adopting these sustainable practices and investing in the right precision farming equipment, farmers can ensure the long-term viability of their operations, reduce input costs, and minimize their environmental footprint.

Choosing Appropriate Tractors and Implements: Maximizing Efficiency

Selecting the right tractors and implements is vital for maximizing the benefits of precision farming. Whether you are purchasing new or used equipment, it is important to ensure that the machinery aligns with your farming goals and can support advanced technologies.

When considering agricultural machinery for sale, keep the following factors in mind:

- **Compatibility with GPS systems:** Make sure the equipment can integrate seamlessly with your existing or planned GPS technology to optimize field operations.
- **Power and size:** The tractor's power should match the needs of your farm and the implements you plan to use, ensuring efficient operation without overloading.
- **Fuel efficiency:** Choose tractors with advanced engine technology for improved fuel economy, which helps lower operating costs over time.
- **Comfort and ergonomics:** Since farming often involves long working hours, consider the operator's comfort. Features such as adjustable seats, climate control, and intuitive controls can improve operator productivity.
- **Smart features:** Many modern tractors come with built-in precision farming capabilities, such as automatic steering and data collection tools that enhance operational efficiency.

By choosing the right equipment, farmers can ensure that their operations are optimized for the latest technological advancements, improving productivity and reducing costs.

Farm Equipment Rentals: A Flexible Option for Farmers

For farmers who are hesitant to make significant investments in new precision farming equipment, renting can be a viable alternative. Renting allows farmers to access cutting-edge technology without committing to large upfront costs, offering several key benefits:

- **Test different types of equipment before purchasing:** Renting allows you to try out various machinery to ensure that it meets your needs before making a long-term investment.
- **Access specialized machinery for seasonal needs:** Certain equipment may only be needed for specific times of the year. Renting allows you to access these specialized tools without having to store them year-round.

- **Reduce upfront costs and maintenance responsibilities:** Renting eliminates the need for large capital expenditures and reduces the ongoing maintenance and repair costs associated with owning equipment.
- **Stay up-to-date with the latest technology:** Renting ensures that farmers have access to the most advanced precision farming tools without the need for constant upgrades.

Many agricultural equipment dealers offer rental options for a wide range of machinery, from basic tractors and implements to specialized precision farming tools. Renting can be a cost-effective and flexible solution for farmers who want to incorporate the latest technology into their operations.

Agricultural Land for Sale: Essential Considerations for Precision Farming

When searching for agricultural land to purchase, it's important to assess whether the property is suitable for implementing precision farming practices. Precision farming relies mostly on technology, including GPS systems and data analytics, which require land that supports these innovations. The following factors should be considered when evaluating agricultural land for sale:

- **Field size and shape:** Regular-shaped fields are easier to manage using GPS-guided equipment, as they ensure more precise control over planting, irrigation, and harvesting. Irregularly shaped fields can complicate machinery movements, reducing efficiency.
- **Soil quality:** Ideal land should have consistent soil types that respond well to precision management techniques. Uniform soil properties allow for better application of fertilizers, irrigation, and other treatments, maximizing crop yield and minimizing waste.
- **Topography:** Gently sloping or flat land is generally preferable for precision farming equipment, as these terrains make it easier for machinery to navigate and apply treatments consistently across the field.
- **Connectivity:** For precision farming to function effectively, reliable cellular or internet coverage is crucial for data transmission. Ensure that the land has access to adequate communication infrastructure to support real-time monitoring and decision-making.
- **Water resources:** Precision farming usually includes advanced irrigation systems that require dependable water sources. Ensure the property has sufficient access to quality water for efficient and sustainable irrigation.

Selecting land that meets these criteria will better position you to adopt precision farming technologies and maximize your farm's potential for productivity and sustainability.

LASER LEVELERS

A Laser Leveler is a device commonly used in the construction industry for surveying purposes. It helps in determining precise alignments and ensuring surfaces are level. This tool can be used to achieve accurate alignment or leveling of a project in various ways, whether it is vertical, horizontal, or at a specific angle

Figure 7.2: Laser Levelers

The acronym LASER stands for Light Amplification by Stimulated Emission of Radiation. A Laser Leveler utilizes a LASER beam generated by an LED (Light Emitting Diode) to accurately detect the required alignment. The LED in the laser leveler is typically compact, producing a fine, narrow laser beam. This allows the user to easily assess the accuracy of the alignment. The beam can be set to be horizontal, vertical, or at any desired angle, depending on the user's specific needs.

Figure 7.3: Laser Levelers

- Laser levelers use a red beam of light to accurately display the required level.
- These devices feature compact LEDs that emit fine beams of light, ensuring enhanced precision.
- Laser levelers are powered by batteries, typically using rechargeable alkaline batteries.
- In the past, laser levelers were primarily used indoors due to the insufficient brightness of their light beams for outdoor tasks. However, modern laser levelers are now designed to emit powerful beams suitable for outdoor projects.
- The latest models of laser levelers come equipped with remote controls, allowing operation from a distance.

Components of a Laser Leveler

A laser leveler consists of a spirit level (or bubble level) and a diode that emits a laser beam. However, different types of laser levelers feature various components. For instance, a self-levelling laser level has a pendulum to align the spirit level bubble, with LEDs mounted on it. In contrast, a rotary laser leveler includes a motor that rotates the device, providing a 360-degree view.

Working of a Laser Leveler

A laser leveler is mounted on a tripod, and the height of the tripod is adjusted based on the user's requirements. The bubble level is aligned with the reference point, and the alignment is determined using the straight or dotted red light beam emitted from the diode. This indicates the perfect alignment or level for the project, which can be horizontal, vertical, or at any specified angle. Laser levelers are also used to determine grading and slopes.

Uses of a Laser Leveler

Laser levelers have versatile applications, both indoors and outdoors.

Indoor Use of Laser Leveler

Since the 1970s, interior designers, plumbers, and electricians have relied on laser levelers. These tools are ideal for aligning pictures and wall hangings, as well as for the installation of plumbing and electrical wiring within construction projects.

Outdoor Use of Laser Leveler

Laser levelers are valuable for land surveying and landscaping. They are essential for erecting structures such as walls, roofs, and terraces, ensuring precise alignment in outdoor environments.

Application of Laser Leveler

Laser levelers have a wide range of applications, including but not limited to:

- Surveying and levelling elevations.
- Constructing walls, ceilings, and other structures.

- Installing and fitting windows and doors.
- Building terraces, overhead tanks, and septic tanks.
- Electrical wiring installations.
- Laying plumbing lines.
- Landscaping and land shaping.
- Determining slopes.
- Levelling layouts and floors.
- Aligning cabinets, cupboards, and other furniture.

Types of Laser Leveler

Self-Levelling Laser Levels

Self-levelling laser levelers do not require manual calibration every time they are used. After an initial calibration with the bubble in the spirit level, subsequent calibrations are automatically performed by the device.

Rotary Laser Levels

This type of laser leveler is equipped with a motor that enables the device to rotate on a base. This rotation provides the user with a full 360-degree view, facilitating horizontal, vertical, grading, slope, and angled alignments with ease.

Horizontal and Vertical Laser Levels

These laser levelers, either self-levelling or manual, are specifically designed for horizontal or vertical alignment tasks.

Manual Laser Levels

This is the most basic type of laser leveler, requiring manual alignment each time it is used. While it is prone to some errors and less accurate than other models, it remains a cost-effective and durable choice for those on a budget.

Advantage of Laser Levelers

Laser levelers offer several significant advantages:

- **Accuracy:** Laser levelers provide high precision with no need for repeated verification.
- **Durability:** These devices are more durable compared to other surveying and levelling tools, especially when used and stored properly, extending their lifespan.
- **Ease of Use:** Laser levelers are simple to operate, reducing the time and effort required for setup.
- **Simple Setup:** They are easy to set up and do not demand much time or energy.
- **Performance:** Laser levelers are robust and function well even in harsh environmental and climatic conditions.

Benefits of Laser Leveler

- Lowers the cultivation time as land levelling is done in one go.
- Saves electricity and reduces greenhouse gas emissions.
- Higher yields lead to higher income of farmers.
- Easy and balanced rotating parts.
- Safety covers on all the moving parts.
- Improves uniformity during crop maturity.
- Reduces seed rate, fertilizers, chemicals, and fuel requirements.
- Low battery consumption
- Resistant to water, dust, and temperature to work in all weather conditions.
- Saves water and electricity during the levelling process.
- Comes with a safety alarm.

ZERO TILL DRILLS

Traditional wheat sowing after paddy harvesting is a challenging and expensive process, requiring 5-7 tilling operations to prepare the seedbed. This delay in sowing affects crop yield and reduces profit margins. This machine allows for direct sowing after paddy harvest, eliminating the need for seedbed preparation. As a result, it saves on diesel, labor, and time, while significantly increasing the yield.

Figure 7.4: Zero Till Drills

Benefits of Zero Till Drills

- Crops can be sown in a larger area in a limited time.
- Saves irrigation water up to 15 per cent.
- Saves operation cost and is eco-friendly.

- Good germination of wheat crop with proper seed distribution.
- Improves soil structure and fertility.

A Step-by-Step Guide for Implementing Precision Farming

Adopting precision farming can initially seem overwhelming, but with the right strategy, the transition can be both smooth and beneficial. Below is a detailed step-by-step guide to help implement precision farming techniques on the farm:

- **Evaluate current operations:** Review existing farming methods and pinpoint areas where precision farming can provide the most significant improvements.
- **Establish clear objectives:** Determine what you aim to achieve with precision farming, such as cutting input costs or enhancing yield.
- **Begin with soil mapping:** Perform an in-depth soil analysis to understand the variation within your fields.
- **Invest in GPS technology:** Start by incorporating basic GPS guidance systems for your tractors and equipment.
- **Apply variable rate technology:** Use the data from your soil map to apply inputs at varying rates across your fields.
- **Adopt yield monitoring:** Equip your harvesting machinery with yield monitors to track crop productivity.
- **Integrate data management systems:** Utilize farm management software to gather and analyze data from multiple sources.
- **Explore advanced technologies:** Gradually incorporate more sophisticated precision farming tools as you become more comfortable with the fundamentals.
- **Commit to continuous learning:** Keep up-to-date with emerging technologies and best practices in precision farming.
- **Review and adapt:** Regularly assess your precision farming methods and adjust your strategies based on outcomes.

Remember, transitioning to precision farming is a continuous process. Begin small, learn from your experiences, and expand your use of technology as you observe the benefits.

Future Vision of Precision Farming

With ongoing technological advancements, the future of precision farming holds great promise. Key emerging trends to watch include:

- **Artificial Intelligence (AI) and Machine Learning:** These technologies will enable deeper data analysis and more informed decision-making in farm management.
- **Internet of Things (IoT):** The enhanced connectivity between farm equipment and sensors will result in more automated and responsive farming systems.
- **Robotics:** Autonomous robots are becoming increasingly capable of handling tasks such as planting, weeding, and harvesting.
- **Vertical farming:** Precision farming methods are being adapted to indoor and urban farming environments.

- **Blockchain in agriculture:** This technology will improve traceability and transparency throughout the food supply chain.

By staying informed about these emerging trends and consistently adapting your farming practices, you can ensure your farm remains competitive and sustainable in the future.

PNEUMATIC PLANTERS

Pneumatic planters are agricultural machines designed to plant seeds efficiently using air pressure to deliver seeds into the soil. These planters rely on a pneumatic (air-driven) system to transport seeds from a seed hopper to a planting mechanism, allowing for precise seed placement and spacing. They are commonly used in large-scale farming operations for crops like corn, soybeans, cotton, and other row crops. Here's a closer look at how pneumatic planters work and their key benefits:

Working of Pneumatic Planters

- **Seed Hopper:** Seeds are stored in a hopper at the top of the planter. This hopper feeds the seeds into the pneumatic system.
- **Air Delivery System**: Air pressure is generated by a fan or blower that pushes the seeds through tubes. The airflow carries the seeds toward the planting units. The air system is typically adjustable to ensure precise seed delivery based on seed size, planting depth, and row spacing.
- **Seed Metering System:** The seed metering mechanism regulates the flow of seeds to maintain the correct planting rate. This can be a disc, vacuum, or other type of metering system that uses air pressure to control seed release. The pneumatic system ensures that seeds are spaced evenly, which improves crop uniformity.
- **Planting Units:** The planting units are where the seeds are actually placed in the soil. These units may have furrow openers to create a groove in the soil, drop tubes to release the seeds, and press wheels to cover the seeds after planting.
- **Precision and Accuracy:** Pneumatic planters are known for their ability to place seeds at precise depths and spacing, which leads to improved crop emergence and growth. The consistent seed placement also minimizes the risk of crowding or gaps in plant growth.

Merits of Pneumatic Planters

- **Precision Planting:** The ability to control seed depth, spacing, and population density leads to more uniform crop emergence. This can significantly improve yields.
- **Minimal Soil Disruption:** The planter's precise seed placement means there is less soil disturbance, which can benefit soil structure and moisture retention.
- **Reduced Seed Damage:** Because the air-driven system is gentle on seeds, there is less physical damage compared to mechanical planters.

- **Flexibility in Seed Types:** Pneumatic planters can handle a wide range of seed types, from small seeds (like canola or soybean) to larger seeds (like corn or cotton), due to their adjustable air delivery systems.
- **Faster Planting:** Pneumatic planters can plant large areas quickly, especially in combination with modern GPS technology and variable rate control systems.

Disadvantages of Pneumatic Planters

- **Complexity and Maintenance:** Pneumatic systems require more maintenance and can be complex compared to mechanical seeders. The air pressure system, hoses, and fans must be kept in good working order.
- **Power Requirements:** Pneumatic planters require a reliable power source to operate the air delivery system, which can put additional strain on tractors or other equipment.
- **Cost:** Pneumatic planters tend to be more expensive than traditional mechanical planters, due to the additional components involved.

Types of Pneumatic Planters

- **Vacuum Planters:** These use a vacuum to hold seeds against a seed plate, releasing them one at a time to ensure accurate spacing.
- **Pressure Planters:** These use positive air pressure to push seeds through tubes and into the soil, offering more control over seed distribution.
- **Double-Disc and Single-Disc Pneumatic Planters:** These refer to the type of disc opener used in the planter units to create furrows for seed placement.

Final Thoughts

Pneumatic planters are highly effective tools for modern farming, offering precision, speed, and flexibility in planting. They are particularly beneficial for large-scale operations that require uniform seed placement and high planting accuracy. However, the initial cost and maintenance requirements are higher compared to more traditional planting methods.

8

Bulldozers

Bulldozers are among the most crucial and versatile machines used in construction, mining, and various earth-moving applications. Known for their immense power and capability to perform heavy-duty tasks, bulldozers have become a cornerstone of modern construction projects. Their design features a large blade at the front, allows them to efficiently push large quantities of soil, sand, rubble, or other materials, making them indispensable in clearing land, grading surfaces, and moving materials across construction sites. Advancements in technology and engineering have significantly improved bulldozer performance over the years adding more precision, power, and efficiency to these machines. The use of bulldozers spans various industries, from construction and road building to agriculture and forestry, where they play a key role in shaping landscapes, preparing sites, and assisting in major earth-moving operations. With the ability to work in diverse terrains and under challenging conditions, bulldozers have proven to be resilient, durable, and highly adaptable, meeting the needs of both small and large-scale projects. This chapter provides a breif overview into the history, design, types, functions, and applications of bulldozers, offering a comprehensive understanding of their significance in the construction and earth-moving industry. Through this exploration, we will also look at the innovations that continue to shape the evolution of this powerful machine.

TYPES OF COMMON BULLDOZERS

There are several different types of bulldozers available in the market. The most common among them are listed below:

- **Crawlers:** This bulldozer variant resembles a tractor and moves across the ground using tracks, providing better traction and preventing it from sinking into softer surfaces like loose dirt or sand. Often equipped with a ripper at the rear, crawlers are capable of breaking up tougher surfaces. Larger crawlers are commonly used in road construction and mining applications, where they transport large volumes of materials. Smaller versions, on the other hand, are used as grading tractors for less demanding earth-moving tasks.

- **Mini Dozers:** The smallest bulldozer models, mini dozers are perfect for working in confined spaces where larger, bulkier machines cannot fit. They offer exceptional maneuverability and versatility, making them well-suited for tasks such as sloping, residential lot clearing, grading, and backfilling.
- **Wheel Dozers:** Unlike crawlers, wheel dozers rely on wheels for movement. These machines are larger, faster, and more maneuverable, making them ideal for high-performance tasks. However, they can struggle on softer terrains, where they may become stuck more easily than crawlers.
- **Shiphold/Port Handling Dozers:** Caterpillar offers a specialized line of dozers designed for efficient maneuvering in and out of cargo holds on commercial vessels. These dozers are also effective at moving various bulk materials in port settings.
- **Mulchers:** A specialized crawler dozer, the mulcher features an attachment for performing tasks related to forestry, construction, and roadwork. This attachment allows for more precise and effective vegetation clearing and land management.

DIFFERENT TYPES OF BULLDOZER BLADES

The bulldozer types mentioned above can be further categorized based on the blades they use. Different blade designs serve distinct purposes, can handle various materials, and manage different load weights. Below are the most common types of blades:

Straight Blades (S-Blade)

Straight blades, also known as S-blades, are among the most common dozer blades. They are ideal for grading projects, particularly for leveling and smoothing fine-grained soil.

The S-blade is the shortest blade available for bulldozers and does not have side wings. It attaches to the arm at the lower back corners of the blade. Due to its shape, the straight blade is best suited for medium- to hard-density materials, including fine-grained soils. However, its straight design limits the dozer's ability to lift and carry materials. Some of the best tasks for S-blades include stumping, grading, backfilling, and evening out soil.

Universal Blade (U-Blade)

The curved universal blade, commonly known as the U-blade, is equipped with large side wings that make it ideal for pushing and containing materials. The curved shape, combined with these side wings, allows the U-blade to efficiently push materials over long distances, preventing them from spilling over the sides. These wings are particularly helpful in controlling material flow during operation. Similar to the S-blade, the U-blade attaches to the lower back corners of the blade.

The U-blade is the largest type of dozer blade in both height and width. It is designed for use with soft to medium-density soils and is well-suited for tasks that involve moving large volumes of material. Some of the most common applications for U-blades include ditching, hauling, pushing, and crowning, making it a versatile and powerful tool for various earth-moving tasks.

S-U (Semi-U) Blade

These blades combine the straight and universal blades' features in a narrow, curved shape with large side wings. They assist with crowning, moving heavy material, stumping, ditching and pushing.

This blade combines features from the S-blade and the U-blade to give it stronger penetration and better overall versatility. It's narrower, less curved and has side wings smaller than a normal U-blade. This design makes it ideal for pushing soil across long distances. This blade attaches in the lower back of the blade using angled stabilizing braces and either one or two hydraulic tilt cylinders. It's best for pushing soft- to medium-density sand and soil. Some of the best tasks for an S-U-blade include crowning, moving heavy material, stumping and ditching.

Angle Blade

When there is a need to move debris or material off to the side, an angle blade will give you the most efficient motion. Angle blades are two-way blades that help with stumping, shaping, stripping and ditching.

This type of blade attaches to the center of the bulldozer's panel. Its location is useful for moving debris to the side since it can angle close to 30 degrees left or right. Due to this, an angle blade is considered a two-way blade. Keep in mind that this blade can spill since it does not have side wings. It's a great choice for projects involving soft- to medium-hard-density soils, snow and gravel. Some of the best tasks for angle blades include stumping, shaping, stripping and ditching.

Power-Angle-Tilt (PAT) Blade

This blade is multidirectional. The operator can control the blade to angle, tilt and lift in almost any direction for scraping, clearing, leveling and grading applications.

The PAT blade is one of the most versatile blades thanks to its easy maneuvering and multifaceted motions. The driver controls the blade from the cabin and can angle, tilt and lift in almost all directions. Like angle blades, these blades are mounted in the center of the panel. It's also similar to the angle blade in that it's best to use with soft- to medium-hard-density soils. Some of the best tasks for PAT blades include scraping, land clearing, leveling, backfilling and grading.

BULLDOZER PARTS AND FUNCTIONS

Bulldozers are versatile machines, designed for various terrains, thanks to their distinct features like blades, rippers, and whether they use tracks or tires.

- **Rippers:** A ripper is a claw-like attachment at the rear of a bulldozer, used to break up land for agricultural purposes or to break down rock and earth.

There are single-shank and multi-shank rippers available, depending on the project's needs.

- **Final Drive:** The final drive is a critical part of the bulldozer, typically the most used and replaced component. It distributes the load across multiple gear teeth and elevates the drive motor away from suspension.
- **Cab:** The cab is where the operator controls the bulldozer, and modern cabs are designed for safety and comfort. Features like noise reduction and impact absorption are important for operators who spend long hours inside.
- **Tracks/Tires:** Tracks provide stability on hard, uneven terrain, while tires are more suited for soft ground. Choosing the right option depends on the nature of the worksite.
- **Engine:** Bulldozers require high-powered engines to move heavy loads. Newer engines often produce fewer emissions to meet EPA standards.
- **Push Frame:** The push frame is essential for positioning materials and helps move the blade.
- **Bulldozer Blades:** The blade at the front of the bulldozer is used to push and dig through materials. Different types of blades are suited for specific tasks.
- **The Bulldozer Undercarriage:** The undercarriage supports the bulldozer, including tracks, rollers, idlers, and sprockets, working together for stability across various terrains.
- **Tracks:** Made of durable steel, tracks distribute the bulldozer's weight evenly, preventing it from sinking into soft or uneven surfaces.
- **Rollers:** Located on the undercarriage, rollers support the bulldozer's weight, ensuring smooth track movement.
- **Idlers:** Positioned at the front and rear of the tracks, idlers help manage track tension and prevent derailment.
- **Sprockets:** Located at the rear of the bulldozer, sprockets use engine power to move the tracks.
- **The Bulldozer Ripper:** The ripper is used to break hard ground, rocks, and materials that cannot be moved by the blade. Its pointed tip digs into the ground, loosening material.

Variants of Ripper Shanks

- **Single Shank Ripper:** This type has one heavy-duty shank for breaking materials like rock and compacted soil.
- **Multi-Shank Ripper:** This version uses multiple shanks for lighter projects, such as breaking asphalt and frozen ground.

Bulldozer Hydraulics

The hydraulic system powers various components, including the blade and ripper. Pressurized fluid is used to provide precise control of the bulldozer's movements.

Key Components of Hydravlics

- **Pump:** The pump, powered by the bulldozer's engine, generates the pressure needed for hydraulic fluid movement.
- **Valves:** These control the flow and direction of the hydraulic fluid.
- **Cylinders:** Cylinders activate the pressurized fluid, driving the movement of attachments.
- **Hoses:** Hydraulic hoses transport the fluid between components.
- **Bulldozer Engine and Transmission:** These components are crucial for providing power and movement. The engine generates horsepower, while the transmission uses this power to maneuver the bulldozer.

Types of Engines

- **Diesel Engines:** Diesel engines are commonly used in bulldozers for their fuel efficiency and high power output.

Transmission Types

- **Hydrostatic Transmission:** Uses hydraulic fluid for precise control of speed and direction.
- **Powershift Transmission:** Uses gears for power transfer and is ideal for heavy-duty tasks.

Other Essential Bulldozer Parts

- **Cooling System:** Ensures the engine remains cool, preventing overheating.
- **Electrical System:** Powers various electrical components like lights and controls.
- **Operator Cab and Controls:** The cab provides a safe working environment and easy access to machine controls.
- **Final Drives:** Transfers power from the transmission to the tracks, enabling smooth operation of the bulldozer.

Choosing the Egiht Bulldozer

Selecting the right bulldozer for your project, whether purchasing or renting, can significantly enhance its efficiency. Consider the following factors when evaluating bulldozer options:

- Determine its primary function. Are you leveling the ground, pushing materials, or performing other tasks?
- Assess the work site. Do you need to operate in confined areas?
- Evaluate the terrain. Will you be working on hard or soft surfaces?
- Identify the materials to be moved. What kind of earth are you handling, and what is its density?

Different bulldozer models can meet most of your hauling, pushing, and leveling needs. By taking the time to assess your options and understand the specific features of each bulldozer, you'll make a more informed decision for your worksite.

Bulldozer Maintenance Tips

Routine maintenance is important for maintaining a bulldozer's performance and extending its lifespan. Here are some tips to keep your bulldozer in peak condition:

General Maintenance Practices

- **Daily Inspections**: Check the fluid levels, inspect the tracks, and look for any signs of wear or damage.
- **Weekly Maintenance:** Lubricate the moving parts, check hydraulic fluid levels, and examine the blade and ripper.
- **Monthly Checks:** Inspect the undercarriage, engine, and transmission for leaks and signs of wear.
- **Annual Service:** Perform oil and filter changes, replace coolant, and conduct a thorough inspection of the five main components.

Significance of Using Genuine Parts and Lubricants

Always use parts and lubricants designed specifically for your bulldozer. Using incorrect products can cause significant damage to your machine. Choosing compatible lubricants and parts will help minimize the risk of engine failure.

Safety Precautions

- **Personal Protective Equipment (PPE):** Always wear gloves, safety glasses, and steel-toed boots when operating or maintaining the bulldozer.
- **Lockout/Tagout Procedures:** Ensure the bulldozer is turned off before beginning any maintenance work.
- **Proper Tools and Training:** Only operate the bulldozer after receiving the proper training and using the correct tools.

TRENCHERS

Trenchers are specialized machines designed for trench excavation, a task that can be highly hazardous. Trenches are essential for numerous construction and utility projects, and trenchers often offer a more efficient alternative to excavators in completing this work.

The history of trenchers is rich and dates back to the time when trench excavation was carried out manually, relying on human labor. These early tools required significant physical effort to dig through the earth, setting the foundation for the development of the heavy machinery we use today on construction sites across the globe.

As years passed trenchers have incorporated technological innovations that have drastically improved their ability to cut through a wide range of terrains with greater precision and efficiency. This section will delve into the components and structure of trenchers, exploring the different types of trenchers, their various applications, safety protocols, and emerging trends in the field.

Various Types of Trenchers

Trenchers are available in a variety of types, each tailored to meet specific excavation needs and project demands. Understanding the differences between these machines is essential for selecting the most suitable one for trenching tasks. The four main types of trenchers are wheel trenchers, chain trenchers, micro trenchers, and portable trenchers. Each type offers distinct advantages for different construction applications. Additionally, trencher attachments can be added to common construction machinery for enhanced functionality.

Mini Excavators

Mini excavators are compact yet highly versatile machines frequently used for trenching. With their powerful hydraulic systems and small footprint, these machines can easily dig straight and uniform trenches in tight spaces. Mini excavators are ideal for residential construction, landscaping, and small utility installation projects. Their narrow profiles allow them to maneuver through restricted spaces with ease, ensuring precise trenching results.

Chain Trenchers: For Deep and Narrow Excavations

Chain trenchers are specifically engineered for creating narrow and deep trenches quickly and effectively. They utilize a large, rotating digging chain with cutting teeth that carve through the soil to form the desired trench shape. Available in various sizes, some chain trenchers can dig trenches several feet deep. These machines are commonly used in utility installation, pipeline construction, and irrigation work.

Wheel Trenchers: The High-Speed Trencher

Wheel trenchers are designed for projects that require clean, straight trenches. These machines feature a wheel-mounted cutting system, allowing them to operate efficiently on hard or compacted surfaces. Their high-speed operation makes them ideal for large-scale projects that demand both precision and speed in trenching.

Skid-Steer Loaders

Skid-steer loaders are commonly used for trenching due to their ability to perform a wide range of tasks with interchangeable attachments. When fitted with a trencher attachment, these loaders can create trenches of various depths and widths with minimal effort. Known for their excellent maneuverability, skid-steer loaders are perfect for working in confined spaces and tight corners. They are often utilized in landscaping, utility installation, and small construction projects.

Micro Trenchers: Revolutionizing Urban Utility Installations

Micro trenchers are compact machines used in urban or residential environments where minimal surface disruption is crucial. These specialized trenchers are primarily used for laying fiber optic cables or small-diameter pipes with minimal environmental impact. Their small size allows them to easily navigate dense urban areas and utility networks, providing a solution for modern infrastructure projects.

Auger Trenchers

Auger trenchers are versatile machines that dig precise trenches with minimal soil disturbance. These machines employ a rotating auger that gradually removes soil to form the trench profile. Auger trenchers are particularly effective in challenging soil conditions such as rocky or compacted earth. They are mostly used in utility installations, foundation work, and pipeline construction.

Vibratory Plows

Vibratory plows are specialized machines that create narrow, shallow trenches by using vibrations to break through the soil. These machines are especially useful for installing cables and pipelines just beneath the surface, without causing significant disturbance to the ground. Vibratory plows are commonly used in utility construction, telecommunication work, and irrigation systems. Their low impact on the landscape makes them a popular choice for trenching in areas where maintenance of the existing environment is important.

Portable Trenchers

Portable trenchers are lightweight, flexible machines designed for small-scale projects, DIY tasks, and working in tight spaces. These machines are user-friendly, making it easy for operators to navigate confined areas with minimal effort. Portable trenchers are especially favored for landscaping projects, residential construction, and similar applications, offering an accessible and efficient solution for a variety of trenching needs.

Operating a Trencher

Operating trenchers involves different steps depending on the type of machine in use. However, the basic steps for both wheel and chain trenchers typically follow these guidelines:

- Adjust the speed and boom height, which are usually controlled by hydraulic levers.
- Engage the trencher's engine to power the machine.
- Activate the trencher's chain or wheel, depending on the model.
- Release the safety lock, usually found on a handle, to enable operation.
- Drive the trencher in the desired direction, adjusting the speed and depth of the boom as necessary to maintain control and precision throughout the excavation.

Common Applications of Trenchers

Trenchers are versatile machines widely used in the construction and utility industries due to their efficiency in trench creation, providing time-saving advantages across a variety of applications. Their ability to rapidly create trenches significantly streamlines the excavation process.

When utility and cable laying projects are needed, trenchers offer an effective method for excavating precise and well-defined channels for electrical cables, water pipes, or telecommunication lines. The use of trenchers in these situations ensures that installations can be carried out with minimal disruption to the surrounding environment, saving both time and resources.

Drainage projects also benefit from the use of trenchers. For both residential and large-scale drainage systems, trenchers facilitate the excavation of trenches for new pipes or infrastructure designed to redirect water flow. This is especially important for preventing flooding and soil erosion, particularly in larger-scale construction or renovation efforts.

In agriculture, trenchers help in establishing efficient irrigation systems. Their rapid operation is essential for building water distribution networks that improve water usage efficiency on farms. By creating trenches with minimal disruption to the soil, trenchers play an important role in promoting sustainable agricultural practices.

In road construction projects, trenchers are often used to install utilities beneath road surfaces. Trenchers prove invaluable in these instances by quickly creating the necessary trenching spaces to lay cables, pipes, or other infrastructure components.

Operating a trencher efficiently not only boosts productivity but also ensures a safe working environment. As trenchers are powerful machines equipped with sharp blades, it is essential to pay close attention to safety protocols to avoid accidents. Below are key safety tips for operating trenchers with efficiency and care:

Proper Training and Familiarization

Before using a trencher, operators should undergo thorough training, ideally with an experienced operator. The training should cover not only the basic operations of the trencher but also safety measures, emergency procedures, and routine maintenance. It is important for operators to be familiar with the machine's features and limitations, ensuring that they can safely handle the equipment and use it to its full potential.

Pre-Operation Inspection

Before operation begins, a comprehensive pre-operation inspection should be conducted to ensure the trencher is functioning properly. This includes checking all components such as the cutting mechanism, engine, conveyor system, and wheels or tracks for any signs of wear or damage. Additionally, ensure that all safety features, including guards and emergency stops, are working properly to protect both the operator and others on the worksite.

Adherence to Safety Protocols

Strict adherence to safety protocols is mandatory when operating a trencher. Operators should always wear the appropriate personal protective equipment

(PPE) such as helmets, gloves, and sturdy footwear. It is also crucial to know the clearance requirements for the machine, ensuring that no unauthorized personnel are within the danger zone during operation. Effective communication between operators and other workers is important to avoid accidents and ensure the smooth progress of the operation.

Mindful Trenching Practices

Operators must plan the trenching route carefully before commencing. This involves considering the type of soil, the required trench depth, and any potential obstacles that may be encountered. To ensure precise and controlled excavation, operators should avoid abrupt movements and maintain a steady, controlled speed throughout the operation. Extra caution should be taken when working in challenging terrains, such as sloped or uneven surfaces, to maintain stability and accuracy.

Emergency Response Preparedness

Despite proper training and adherence to safety measures, unexpected situations may arise. Therefore, operators should be well-prepared to handle emergencies by knowing how to shut down the trencher safely in case of a malfunction. Emergency contact information and procedures should be easily accessible, and all workers on-site should be familiar with the emergency response plan to ensure swift action in case of an accident.

By emphasizing training, conducting thorough inspections, following safety protocols, practicing mindful trenching, and preparing for emergency situations, operators can ensure both high efficiency and safety while operating trenchers, minimizing the risk of accidents on the job site.

9

Transplanting Equipment

A transplanting machine is a vital agricultural tool used to move seedlings from a nursery to a field, a process known as transplanting. These machines are specifically designed to plant seedlings or young plants at the desired depth, spacing, and arrangement in the soil, improving the efficiency and precision of transplanting. By replacing manual labor, transplanting machines save time, increase consistency, and ensure higher productivity. These machines are especially useful for crops such as rice, vegetables, and flowers, as they enable faster and more accurate planting compared to traditional methods.

VARIOUS TYPES OF TRANSPLANTING EQUIPMENT

Transplanters are versatile machines used for planting a wide range of crops, including:

- Tomatoes
- Cabbages
- Strawberries
- Tobacco
- Rice
- Vegetables like leeks, onions, and salads

These machines are designed to efficiently plant seedlings in rows at precise intervals, ensuring uniform growth. The construction of transplanting equipment varies depending on the type of crop being planted and the specific requirements of the field.

Structure and Design of Transplanting Equipment

- **Frame and Chassis:** The main structure of a transplanting machine is the frame or chassis, which supports all other components. It is made from durable materials such as steel or aluminum to ensure strength and longevity. The frame is designed to be stable and resistant to wear and tear from continuous use in agricultural fields.

- **Seedling Holding System:** The seedling holding system consists of trays or bins where seedlings are placed before being transplanted. These systems are designed to gently hold the seedlings, preventing damage during transportation to the field. Some machines feature conveyor belts or other automated systems to move the seedlings from the holding trays to the planting mechanism.
- **Planting Mechanism:** The planting mechanism is responsible for placing the seedlings into the soil. It generally includes a set of planting forks, wheels, or spades that push the seedlings into the soil at the correct depth and spacing. This mechanism can be adjusted to accommodate different plant sizes and field conditions.
- **Adjustable Depth and Spacing Mechanism:** To ensure precise planting, transplanting machines are equipped with mechanisms that allow the operator to adjust the planting depth and spacing between seedlings. This is crucial for optimizing growth conditions and ensuring that plants have adequate room to develop.
- **Power System:** Transplanting machines can be powered either manually, through a tractor, or by internal combustion engines. Most modern transplanting machines are designed to be compatible with tractors, allowing for efficient operation over larger areas. Some machines are also equipped with electric motors for more precise, smaller-scale operations.

Adjustment of Transplanting Equipment

- **Depth Adjustment:** One of the most critical adjustments on transplanting equipment is the planting depth. It ensures that seedlings are placed at the optimal depth for growth, allowing the roots to establish themselves in the soil. Depth adjustment is often done by altering the height of the planting mechanism, which can be done manually or with hydraulic controls.
- **Spacing Adjustment:** Proper spacing between seedlings is crucial for maximizing crop yields. Many transplanting machines allow operators to adjust the spacing between plants, which is typically done by modifying the planting mechanism or using interchangeable planting discs or wheels. Adjustable spacings ensure that plants have adequate space to grow without competing for resources.
- **Seedling Size Adjustment:** Different crops and seedlings require different levels of precision when it comes to planting depth and spacing. Modern transplanting machines often have adjustable mechanisms to accommodate various seedling sizes. For example, smaller seedlings may require a shallower planting depth, while larger plants may need a deeper hole.

Working of Rice and Vegetable Transplanters

- **Rice Transplanters:** Rice transplanters are specialized machines designed to plant rice seedlings in flooded fields, known as paddy fields. These machines typically consist of a set of rollers or wheels that move the seedlings from trays or bins to the flooded soil. The seedlings are then placed in rows with precise depth and spacing. Rice transplanters can be operated using a walking or tractor-mounted system, making them adaptable to different field sizes. The key features of rice transplanters include their ability to plant in submerged conditions and adjust the row spacing for efficient irrigation and growth.
- **Vegetable Transplanters:** Vegetable transplanters are designed for use in dry or irrigated fields, where seedlings are planted at specific depths and spacing to optimize vegetable growth. These machines often feature multiple rows of planting mechanisms to plant various vegetable species simultaneously. The machine's design allows for high-speed operation, making it ideal for large-scale vegetable farming. Some vegetable transplanters are designed to handle a variety of crops, such as tomatoes, cabbage, and lettuce, and come with interchangeable planting tools to cater to different plant types.

CRCIAL FEATURES OF TRANSPLANTING MACHINES

- **Seedling Handling:** The machine typically picks up seedlings from trays or nurseries and places them in the field.
- **Precise Planting Depth:** It ensures that each seedling is planted at the correct depth to promote optimal growth.
- **Row Spacing:** The machine can adjust the spacing between rows, which is crucial for crop growth and maximizing yields.
- **Speed and Efficiency:** Mechanized transplanting can significantly reduce the time and labor required compared to manual transplanting.
- **Adaptability:** Some machines are versatile and can be adjusted for various crops, like rice, tomatoes, and lettuce.

What is a transplanter used for and where it is used in particular?

Transplanters are typically used for vegetables (tomatoes, leeks, salads, onions, tobacco, strawberries, etc.), whose seedlings are grown in greenhouses inside trays (they can be both bare root and with a plate of different shapes, cubic, pyramidal, conical, cylindrical).

It is, hence, clear that the use of transplanters allows for a substantial reduction in the manpower for planting seedlings.

COMPONENTS OF A TRANSPLANTER

Semi-Automatic Transplanter with Rotating Distributor

This is the most common type of transplanter, consisting of a frame with one or more transplant elements fixed at a distance corresponding to the desired row spacing. Each transplant element includes a seat for the operator, whose job is to transfer seedlings from the trays to the rotating distributor. The core component of the transplanting system is the rotating distributor, which is typically a cylinder with several openings, or cups, where the operator places the seedlings after carefully selecting them. This device, protected and integrated with the share, uses an alternating motion to push the plant into the furrow created by the share itself. The speed at which the distributor rotates depends on the tractor's speed and the desired spacing of the seedlings.

Automatic Transplanters

Automatic transplanters are more complex systems that require fewer operators. These machines only need one operator, who loads the plant trays and guides the transplanting elements. The machine then automatically extracts the seedlings using a system of ejectors, while a set of pliers or discs deposits the seedlings in the soil.

Distributor Handling

In traditional (non-electric) transplanters, the movement of the distributor is powered by a mechanical transmission system, which uses chain and pinions to transfer the rotation from one or more wheels that are in contact with the ground. The transplanting distance, which varies depending on the type of plant, is adjusted by modifying the gear ratio between the ground wheel and the distributor shaft.

Auxiliary Devices

Many transplanters generally include systems for distributing fertilizers and micro granules essential for plant growth. These are typically volumetric distributors, either screw or vane types, designed for solid granular products. In some cases, these devices are combined with localized watering systems that synchronize with the plant.

RICE TRANSPLANTERS

Rice is a staple food for over half of the world's population, but cultivating it can be labor-intensive and time-consuming. To alleviate this challenge and enhance efficiency, rice transplanters have been developed. These machines assist in transferring rice seedlings from nurseries to fields, ensuring consistent planting and reducing manual labor. Machine transplanting involves using a machine to plant young rice seedlings into puddled soil.

Rice transplanters are typically categorized by their operation mode and transplanting speed. In terms of operation mode, they are divided into walking-type and riding-type rice transplanters. Based on transplanting speed, they are classified into standard rice transplanters and high-speed rice transplanters. Walking-type transplanters are always of the standard variety, while riding-type transplanters can be either standard or high-speed models.

The components of rice transplanters are usually the same across different models: a manual rice transplanter includes a seedling box, a transplanting mechanism, a frame, and a floating body (boat board), along with other seedling-related components.

There are two primary types of rice transplanters: the walk-behind type and the riding type.

Operating Principle of Rice Transplanter

The operation of a rice transplanter is designed to automate the traditional manual method of transplanting rice seedlings. It works through a mechanism involving three main stages: seedling pickup, seedling delivery, and planting.

Seedling Pickup

The process begins with the machine picking up the seedlings from the nursery beds using specialized pickup discs or tines. These components gently lift the seedlings, along with a small amount of soil, without damaging the roots.

Seedling Delivery

Once picked, the seedlings are transferred to a conveyor system or seedling tray, which holds them in place while they are moved to the planting area. This ensures consistent spacing and prepares the seedlings for precise planting.

Planting

As the transplanter moves forward, it plants the seedlings at the predetermined spacing and depth in the paddy field. A plunger or opener creates a hole in the soil, and a vacuum system or mechanical grippers place the seedlings into the hole. The closing wheel then covers the seedling with soil, securing it in place.

Key Components of Rice Transplanter and Their Function

Pickup Unit

This component is responsible for lifting the seedlings from the nursery, using rotating discs or tines that work beneath the soil surface to minimize root damage.

Conveyor System/Seeder Tray

This part temporarily holds the seedlings before planting, ensuring uniform spacing and preventing overcrowding or gaps.

Planting Unit

The planting unit includes plungers or openers that create holes in the soil for the seedlings, followed by a vacuum system or mechanical grippers that place the seedlings accurately into the holes.

Closing Wheel

After planting, the closing wheel presses the soil over the base of the seedling, ensuring firm anchorage and proper root-to-soil contact.

Drive System

The drive system powers the transplanter and its various components, often driven by a diesel engine or a Power Take-Off (PTO) connection from a tractor.

Steering and Control Systems

These systems allow the operator to control the direction, speed, and depth of the transplanting operation, adjusting to field conditions.

10

Seedling Raising Technique for Transplanters

The success of mechanized transplanting largely depends on the quality of seedlings used, making seedling raising an essential component of the process. This chapter revolves around the techniques and practices involved in raising seedlings that are optimized for transplanters. Proper seedling raising ensures that the plants are healthy, uniform, and robust, thereby improving the efficiency of the transplanting process and enhancing crop yields. We will explore various methods for preparing seedlings, including the use of specialized nurseries, seedling trays, and appropriate handling techniques to prevent damage. Understanding these techniques is important for achieving optimal performance from transplanters, reducing labor costs, and promoting sustainable agricultural practices.

Transplanting is a common method for controlling weeds in wet or puddled fields. It uses less seed than direct seeding but requires significantly more labor. Additionally, transplanted crops tend to take longer to mature due to transplanting shock.

TECHNIQUES OF MANUAL TRANSPLANTING

Manual transplanting can be performed using two different methods:

Random Method

In the random method, seedlings are transplanted without maintaining a set distance or spacing between plants. While it's difficult to establish equal distances between each hill in this method, it is crucial that the estimated distances are neither too close nor too wide. The spacing should range between 10 cm and 25 cm to ensure effective weed control through early shading from the rice canopies.

Straight-row Method

This method involves transplanting seedlings in a uniform pattern with consistent spacing. Proper plant spacing is a key factor in transplanting rice, as it can increase yields by 25–40% compared to improper spacing. It also helps reduce costs on inputs, labor, and materials. Straight rows simplify management practices such as hand or rotary weeding and the application of fertilizers, herbicides, or insecticides. This method allows for optimal plant spacing.

In the straight-row method, transplant 2–3 seedlings of 15–21 day-old wet-bed or dry-bed grown seedlings at a spacing of 20 x 20 cm. Closer spacing, such as 15 x 15 cm or 10 x 10 cm, may be used depending on the availability of planters and transplanting costs. This is particularly beneficial when weed control is insufficient.

PRINCIPLES OF CROP ESTABLISHMENT

General Goal

- Establishing a uniform plant stand.
- For maximum yield, plant stand must be uniform across the field.
- Transplanting gives a good stand establishment
- Direct seeding, especially wet direct seeding, may be problematic.

  If direct seed then
  - Need more uniform land leveling and better water management
  - Seed planted too deep results in poor emergence and weak plants

System of Rice Intensification (SRI)

Seed Rate

- 7-8 kg/ha for transplanting a single seedling per hill.
- 12-15 kg/ha for transplanting two seedlings per hill in areas where rice establishment is challenging.
- Seedlings that are 15 days old are transplanted with 1-2 seedlings per hill.
- Transplant the seedlings within 30 minutes of pulling them from the nursery.

  The commonly practiced methods are:

1. Usage of rope with marking
2. Usage of aluminium pipes with markings.
3. Usage of bamboo plank with pegs as markings
4. Usage of wooden marker
5. Usage of roller markers
6. Fill up the gaps between 7th and 10th DAT.
7. There may be difficulty in crop establishment in areas with heavy downpour (North East Monsoon periods of Tamil Nadu).

TRANSPLANTING

Transplanting involves growing seedlings in a nursery and then transferring them to a prepared main field. This method guarantees a consistent plant stand and provides the rice crop with an early advantage over competing weeds.

In Tungabhadra command area, for transplanting of long duration varieties such as sona masuri under delayed condition (August 15-30th) select the seedlings which are 45 days old and transplant about 3 to 4 seedlings per hill to get higher yield.

Planting Seedling in the Main Field

Number of seedlings /hill

- 2 to 4 seedlings / hill
- Saline soil : 4 to 6 seedling / hill

Depth of Planting

- Clay soil : 5 to 6 cm
- Shallow soil : 2.5 to 3.0 cm
- Shallow planting (3 cm) ensures quick establishment and more tillers.
- Deeper planting (> 5cm) leads to delayed establishment and reduced tillers.

Transplanting Shock

It occurs when the seedlings are pulled out from the nursery and planted in the new environment. For recovery from shock, it will take minimum of 5 to 7 days under tropics.

- Shallow planting reduces the period.
- Mild temperature after transplanting also reduces the period.
- Hot and very cold weather delays recovery.
- Best temperature: <30° C maximum and > 20° C minimum.

Gap filling

Fill the gaps, if any within 7 - 10 days after transplanting.

SEEDLING METHODS FOR TRANSPLANTERS

Seedling Raising Techniques for Transplanters are crucial for ensuring healthy, vigorous plants that can be successfully transplanted into the field. Transplanting involves growing seedlings in a controlled environment before they are moved to the field, and the success of the process depends heavily on the quality of the seedlings. Below is a breakdown of the steps and best practices for raising seedlings that will be used with transplanters:

Crucial Factors in Seedling Raising for Transplanters

1. **Selecting the Right Seeds**
    - Use high-quality, certified seeds with good germination rates.
    - Choose varieties suited for the local climate and soil conditions.
    - Ensure seeds are free from diseases, pests, or genetic defects.

2. **Nursery Setup**
    - The nursery can be in a greenhouse, shade net house, or even an open field, depending on the crops and local conditions.
    - For best results, ensure the nursery provides protection from harsh weather conditions, pests, and diseases.
3. **Seedling Media**
    - **Soil Mix:** Use a well-draining and nutrient-rich seedling media, such as a mix of compost, soil, and perlite or vermiculite. A balanced pH (5.5–6.5) is ideal for most crops.
    - **Cocopeat or Peat Moss:** Many seedling trays use cocopeat or peat moss because of their excellent water retention and aeration properties.
    - **Soilless Mix:** Some professional nurseries use soilless mixes that are sterilized to prevent soil-borne diseases.
4. **Germination Conditions**
    - **Temperature:** Maintain optimal temperatures based on the crop. For example, most vegetable seeds germinate best at temperatures between 20°C to 25°C (68°F to 77°F).
    - **Humidity:** Keep humidity high during germination (80-90%) to prevent the seedlings from drying out.
    - **Light:** Some seeds need light to germinate (like lettuce), while others (like tomatoes) need darkness. For most crops, once seeds germinate, ensure they receive sufficient light to prevent leggy growth. If you're growing in a greenhouse or shade house, supplemental lighting may be necessary.
5. **Watering**
    - **Consistent Moisture:** Keep the growing media moist but not waterlogged. Over-watering can lead to root rot and fungal diseases, while under-watering can cause poor germination and weak seedlings.
    - **Watering Methods:** Water gently, either using a fine mist or drip irrigation system to avoid disturbing young seedlings.
6. **Fertilization**
    - **Nutrient Supply:** Seedlings require nutrients, but the amount and timing depend on the growth stage and crop type. Use a balanced liquid fertilizer or slow-release fertilizers for better control.
    - **Timing:** Begin fertilizing when seedlings have developed their first set of true leaves. Fertilization should be light to avoid burning the delicate seedlings.
7. **Thinning and Pruning**
    - **Thinning:** Remove excess seedlings to avoid overcrowding and ensure enough space for each plant. Crowded conditions can lead to weak, spindly plants.

- **Pruning:** Some crops benefit from occasional pruning to promote strong root development and prevent diseases.

8. Acclimatization (Hardening Off)

- Before transplanting seedlings into the field, it's crucial to acclimatize them to outdoor conditions. This process, known as **hardening off**, reduces transplant shock and helps seedlings adapt to environmental stress.
- Gradually expose seedlings to outdoor conditions by increasing their exposure to sunlight, wind, and fluctuating temperatures over a period of 7–10 days.

9. Transplanting Preparation

- **Timing:** Ensure that seedlings are ready for transplanting when they are sufficiently strong but not overgrown. Ideally, seedlings should have well-developed root systems and at least a couple of true leaves.
- **Transplant Spacing:** Maintain proper spacing between seedlings in the nursery to avoid root competition and allow them to grow robustly.
- **Root Pruning:** If necessary, carefully prune or cut the roots to stimulate lateral root growth, which will help seedlings develop better once planted in the field.

Guidelines for Seedling Raising for Transplanters

1. Prepare Seedling Trays or Pots
    - Use seedling trays, pots, or plug trays. These containers should have adequate drainage and are often filled with the chosen growing medium.
2. Sow the Seeds
    - Plant seeds according to the depth recommended for the specific crop. Generally, larger seeds are planted deeper than smaller ones.
    - Ensure proper seed spacing to avoid overcrowding.
3. Maintain Optimal Growing Conditions
    - **Temperature:** Keep the growing area at a consistent temperature suited for seed germination (e.g., 18°C–25°C).
    - **Humidity:** Use misting systems or humidity domes to maintain high humidity, particularly during germination.
    - **Light:** Ensure seedlings get adequate light once they germinate. If using artificial lighting, maintain a light intensity of 1,000–2,000 lux.
4. Water Regularly
    - Water the seedlings regularly to keep the soil moist but not soggy. Water early in the day to reduce fungal issues.
    - As seedlings grow, reduce misting and water with a drip system or watering can to target the roots directly.
5. Fertilize as Needed
    - Start fertilizing seedlings when they have a couple of sets of leaves. Use a liquid fertilizer or a diluted solution of balanced fertilizer.

- Avoid over-fertilization, which can lead to excessive vegetative growth and poor root development.

6. Hardening Off
   - About 7–10 days before transplanting, begin exposing seedlings to outdoor conditions. Gradually increase exposure to full sun, wind, and cooler temperatures to prepare them for the field.
7. Transplant into the Field
   - On the transplanting day, ensure that the soil in the field is well-prepared and ready to receive the seedlings.
   - Transplant seedlings early in the morning or late in the evening to minimize transplant shock.

11

Construction and Function of Irrigation Equipment

INTRODUCTION

Irrigation equipment plays a crucial role in modern agriculture, ensuring that crops receive an adequate supply of water for optimal growth. These systems have evolved over centuries to meet the growing demands of agricultural production, especially in regions where rainfall is inconsistent or insufficient. The construction of irrigation equipment involves various technologies designed to efficiently distribute water to the land, thereby supporting the health and yield of plants. Different types of irrigation equipment are used depending on factors such as the size of the farm, the type of crops being grown, and the availability of water sources.

The most common types of irrigation systems include surface irrigation, sprinkler systems, and drip irrigation. Surface irrigation involves the flow of water across the soil's surface, often through furrows or basins. Sprinkler systems simulate rainfall by spraying water in a controlled manner, while drip irrigation delivers water directly to the roots of plants through a network of tubes and emitters. Each of these systems is designed to optimize water usage, minimize waste, and increase agricultural efficiency.

The function of irrigation equipment goes beyond merely supplying water; it is designed to do so in a way that prevents waterlogging, erosion, and evaporation loss. Properly constructed systems ensure uniform water distribution, reducing the chances of crop damage from over-irrigation or drought stress. In addition to providing water, modern irrigation technologies often incorporate automation and sensors, allowing farmers to monitor and adjust water usage based on real-time conditions. This precision farming approach maximizes resource efficiency, contributes to sustainable agriculture, and helps ensure a stable food supply in regions with challenging climates. As such, irrigation equipment is not just a tool for watering crops, but an integral part of modern agricultural practices that fosters food security.

PUMPS

A pump is a mechanical device designed to move fluids from one location to another. It functions as a hydraulic device, raising a fluid from a lower level to a higher one and transporting it from a low-pressure zone to a high-pressure zone. Pumps work by transforming the mechanical energy of the fluid into pressure energy (hydraulic energy).

Classification of Pumps

Pumps are divided into two main categories: dynamic and positive displacement (aka displacement)

- Dynamic Pumps
    - Centrifugal pumps
    - Vertical centrifugal pumps
    - Horizontal centrifugal pumps
    - Submersible pumps
    - Fire hydrant system
- Positive displacement pumps
    - Diaphragm pumps
    - Gear pumps
    - Peristaltic pumps
    - Cam pumps
    - Piston pumps

Dynamic Pumps

Dynamic pumps come in various types, including centrifugal pumps, vertical centrifugal pumps, horizontal centrifugal pumps, submersible pumps, and fire hydrant systems, which will be discussed below.

Centrifugal Pumps

Centrifugal pumps are among the most widely used pumps globally. Their operation is straightforward, well-documented, and extensively tested. These pumps are known for being durable, efficient, and relatively inexpensive to produce. When in operation, the fluid pressure increases from the pump's inlet to its outlet, with this pressure change propelling the fluid throughout the system.

This pump generates increased force by transferring the motor's mechanical power to the fluid via the rotating impeller. The fluid then enters the impeller's center and is expelled through its vanes. The centrifugal force then boosts the fluid's velocity, converting kinetic energy into force.

Figure 11.1: Centrifugal Pumps

Centrifugal pumps can be classified into three subtypes based on the water flow they produce. The flow pattern is influenced by the shape of the impeller and the design of the pump.

Table 11.1: Centrifugal Pumps Sub-Type and their Description.

Sub-type	Description	Performance
Axial Flow Pump	Also known as a propeller pump, it produces water flow along the impeller shaft direction.	High flow rate Low pressure
Radial Flow Pump	This pump type produces flow in a direction perpendicular to the shaft (90° angle).	Low flow rate High pressure
Mixed Flow Pump	This pump type combines radial and axial flow, producing a conical flow pattern around the shaft.	Medium flow rate Medium pressure

Vertical Centrifugal Pumps

Vertical centrifugal pumps, also referred to as cantilever pumps, feature a distinctive shaft and maintenance design that allows the pump's volume to be directed into the pit, as the bearings are positioned outside the pit. Unlike other models, this pump does not utilize a filled container to cover the shaft but instead uses a throttle bushing. These pumps are commonly used in parts washing applications.

Figure 11.2: Vertical Centrifugal Pumps

Horizontal Centrifugal Pumps

These types of pumps include at least two otherwise more impellers. These pumps are used for pumping services. Each stage is basically a manifold pump.

All stages are housed in the same bunker and mounted on a common shaft. Up to eight additional stages can be installed on separate horizontal shafts. Each stage increases the head by approximately the same amount. A multi-stage pump can also function as a single-stage pump or a double suction pump on the first impeller. Various pumps have been provided and serviced for this type of centrifugal pump.

Figure 11.3: Horizontal Centrifugal Pumps

Submersible Pumps

These pumps are also known as stormwater, sewage and septic pumps. Applications for these pumps include primarily building services, domestic, industrial, commercial, rural, municipal and stormwater recycling applications.

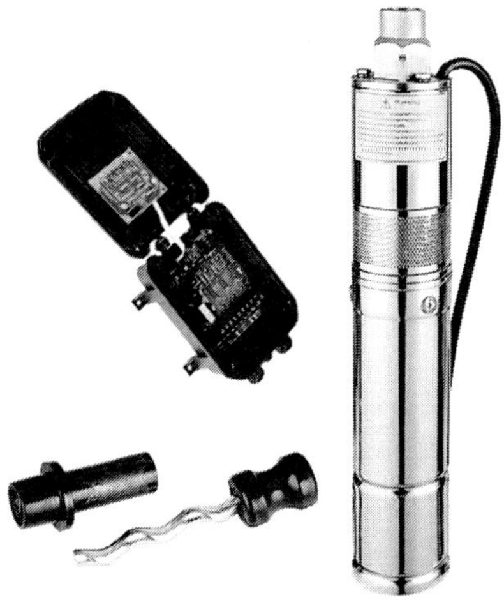

Figure 11.4: Submersible Pumps

These pumps are suitable for transferring stormwater, groundwater, sewage, blackwater, greywater, rainwater, trade waste, chemicals, bore water and food. These plumbing applications mainly include different impellers such as closed type pumps, convection pumps, vortex pumps, multistage pumps, single channel pumps, cutting pumps or grinder pumps. For different applications, a wide range of options are available, including high flow, low flow, low head or high head.

Fire Hydrant System

Fire hydrant pump systems are also known as fire hydrant booster, fire pump and fire pump. These are high pressure pumps designed to increase the firefighting capacity of a building by increasing the force within the hydrant service, as the mains are not sufficient. Applications for this system include mainly irrigation and water distribution.

Figure 11.5: Fire Hydrant System

Positive Displacement Pumps

There are different types of volumetric pumps, some of which will be discussed below, such as diaphragm pumps, gear pumps, peristaltic pumps, lobe pumps, and piston pumps.

Diaphragm Pumps

Diaphragm pumps are also known as AOD pumps (Air Operated Diaphragm), pneumatic and AODD pumps. Applications for these pumps include mainly in continuous applications such as general plant, industrial and mining. AOD pumps are particularly suitable where power is not available and are otherwise used in unstable and combustible areas. These pumps are also used for transferring chemicals, industrial wastewater, food manufacturing, underground coal mines, etc.

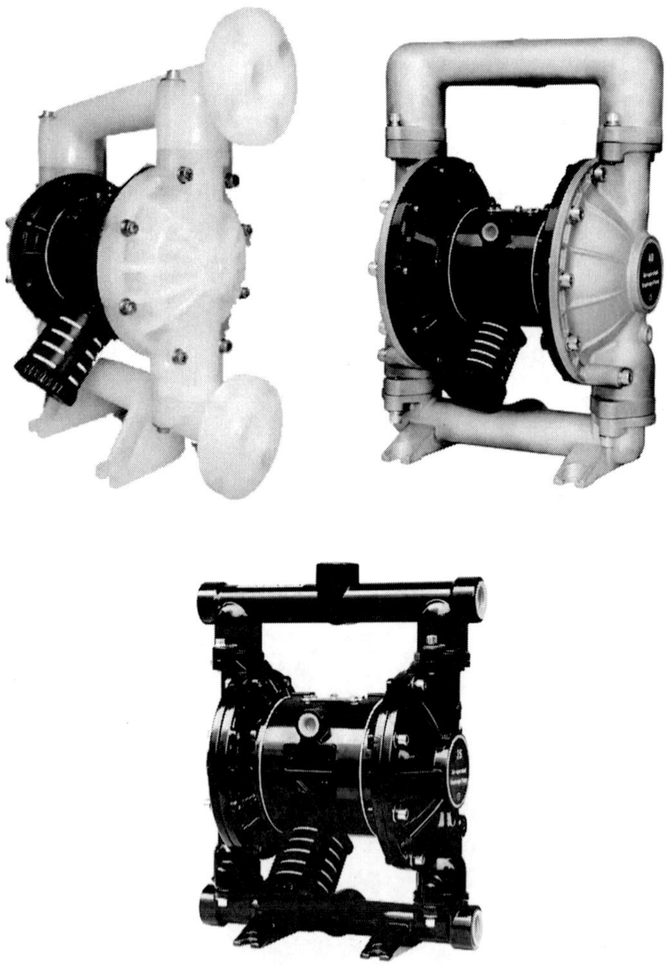

Figure 11.6: Diaphragm Pumps

These pumps are response pumps that include two diaphragms driven by condensing air. The air portion of the transfer valve alternately applies air to both diaphragms; each diaphragm contains a set of ball valves or check valves.

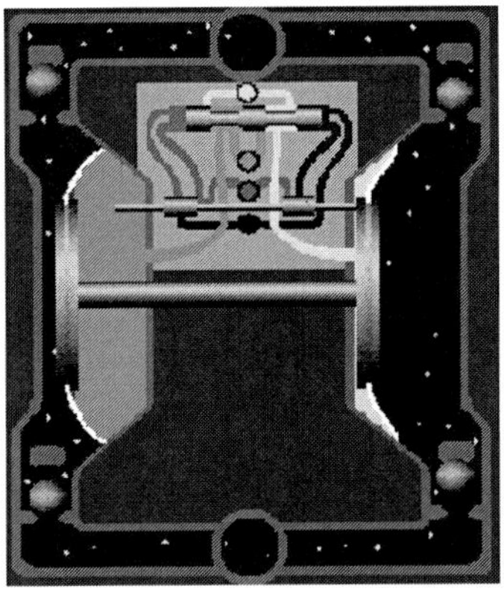

Figure 11.7: Inside Structure of a Diaphragm Pumps

Gear Pumps

These pumps are a rotating positive displacement pump, which means they produce a steady amount of fluid with each revolution. These pumps move the fluid by entering machinery inside and outside the network for a non-exciting pumping action. These pumps are capable of pumping at high forces and can efficiently pump highly concentrated fluids.

Gear pumps do not contain any valves that cause losses such as friction and high impeller speeds. Therefore, the pumps are suitable for handling thick fluids such as fuels and grease. These pumps are not suitable for driving solids and harsh liquids.

Peristaltic Pumps

Peristaltic pumps are also known as tube pumps and peristaltic dosing pumps. It is a volumetric pump and the applications of these pumps are mainly related to processing in the chemical, food and water treatment industries. It provides a steady flow for measuring and mixing and is also capable of pumping various liquids such as toothpaste and various chemicals.

Figure 11.7: Peristaltic Pumps

Lobe Pumps

Lobe pumps are also called rotor pumps and these pumps have different characteristics such as excellent efficiency, rust resistance, sanitary quality, reliability etc. These pumps can handle highly concentrated fluids and solids without harming them. These pumps may work with gear pumps, except for the vanes that do not touch each other. In addition, these pumps have superior pumping chamber compared to gear pumps that allow them to move slurry. These are made of stainless steel and are very polished.

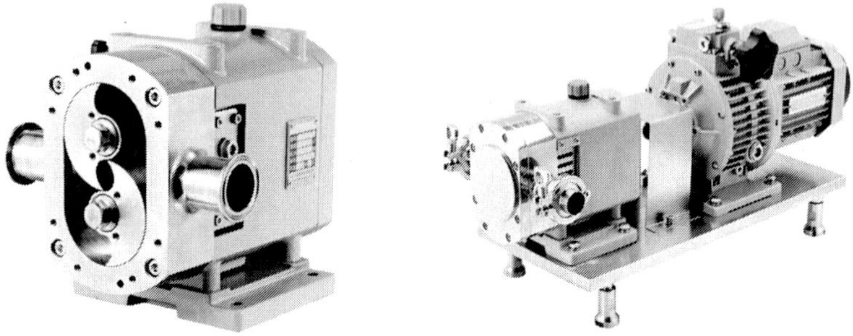

Figure 11.8: Lobe Pumps

Piston Pumps

Piston pump is a positive displacement pump, also known as a plunger pump, in which a high pressure seal responds through a piston. These pumps are often used for water irrigation, scenarios requiring high, reliable pressure and delivery systems for conveying chocolate, pastry, paint, etc.

Figure 11.9: Piston Pumps

Factors to Consider Before Buying a Pump

Farm operations such as irrigation are essential aspects of farming. You need to deliver water to the plants at the right place and time. Agricultural pumps are effective machinery required for running your farm operations. You need an efficient water pump that meets your requirements and enhances your water distribution system to streamline your irrigation operations. Some can be operated using diesel and petrol while others use electricity.

Here are various factors that determine the type of pump that you should choose:

- Size of the field
- Landscape
- Power sources
- Water sources
- Water pressure required
- Engine and fuel consumption
- Displacement and speed of the pump
- Distance and elevation from the water source
- Type of irrigation system that you will be using

When buying a pump, you need something that can operate in streams, ponds, ditches, and canals without priming or suction lines. It should be able to filter out any materials that may clog the nozzles and sprinklers feeding the pivot systems. Besides, it should be easy to use, long-lasting, corrosion-free, and portable.

12

Construction and Function of Harvesting Equipment

Root harvesting equipment is useful in harvesting root crops such as potatoes, carrots, beets, turnips, onions, sweet potatoes, and other underground crops. These machines are designed to dig, lift, and separate the roots from the soil while minimizing damage to the crops. Root harvesting equipment can vary in size and design depending on the crop being harvested and the scale of the operation. Here's an overview of the different types of root harvesting equipment and their features:

VARIANTS OF ROOT HARVESTING EQUIPMENT

Potato Harvesters

- **Two-Row Potato Harvesters:** These machines are designed to harvest two rows of potatoes at a time. They dig up the potatoes from the soil, separate the soil, and lift the potatoes onto a conveyor system for sorting and further processing.
- **Single-Row Potato Harvesters:** These are typically smaller machines used for smaller fields or specialized crops. They work similarly to two-row harvesters but are designed for single-row efficiency.
- **Self-Propelled Potato Harvesters:** These are large, fully mechanized machines capable of harvesting large fields. They often come with additional features like sorting tables and storage bins.
- **Trailed Potato Harvesters:** These harvesters are attached to a tractor and are commonly used for medium to large-scale potato operations. They consist of digging units, separation mechanisms, and conveyors to lift and transport potatoes.

Components of Potato Harvester

- **Digger:** A set of digging blades or shovels that loosen the soil and bring the potatoes to the surface.
- **Separator/Shake:** A vibrating or shaking mechanism that removes excess soil and debris from the harvested crop.
- **Conveyor System:** A belt or conveyor system that transports the potatoes to sorting bins.
- **Sorting Table:** A place where workers can manually sort out damaged or undesirable potatoes.

Carrot Harvesters

- **Single-Row Carrot Harvesters:** These harvesters are used for lifting carrots from the soil one row at a time. They use a digging blade or a set of tines to loosen the soil, followed by a conveyor system to bring the carrots to the surface for cleaning and sorting.
- **Multi-Row Carrot Harvesters:** These machines can handle multiple rows of carrots simultaneously, making them suitable for large-scale operations.
- **Self-Propelled Carrot Harvesters:** Similar to potato harvesters, these machines are self-powered and designed for large-scale commercial operations. They typically come with adjustable digging depths to adapt to different soil conditions.

Components of Carrot Harvesters

- **Digger or Lifter:** A system that loosens the soil around the carrots, making it easier for the crop to be pulled up.
- **Belt or Conveyor:** Carries the carrots from the digging area to the sorting and cleaning system.
- **Cleaning Mechanism:** Removes dirt and debris from the harvested carrots, which may include brushes or high-pressure water jets.

Beet Harvesters

- **Self-Propelled Beet Harvesters:** These harvesters are large machines that dig, lift, and clean beets in one operation. They use a digging share or blade to loosen the soil, followed by a series of conveyors to collect and transport the beets.
- **Tractor-Pulled Beet Harvesters:** Smaller machines that are hitched to a tractor. They are ideal for smaller-scale operations and are used in much the same way as other root crop harvesters.

Components of Beet Harvesters

- **Digger or Blade:** Loosens the soil and pulls the beets out of the ground.

- **Cleaning and Conveying System:** Removes excess dirt and debris and transports the beets to storage bins.
- **Sorting Table:** Used for sorting and removing damaged or undersized beets.

Onion Harvesters

- **Onion Pullers or Lifters:** These machines are designed specifically to pull onions from the soil. They are typically smaller and can be attached to a tractor for field use.
- **Self-Propelled Onion Harvesters:** These harvesters dig and lift onions from the soil, sometimes incorporating shaking or vibrating mechanisms to remove soil and dirt before transporting them to storage bins.

Components of Onion Harvesters

- **Lifting Mechanism:** Often a set of rotating blades or forks that loosen the onions from the soil.
- **Shaker or Separator:** A system that shakes off excess soil and roots.
- **Conveyor and Sorting:** Transports the onions to storage bins, where they are often sorted for quality.

Multi-Crop Root Harvesters

- **Universal Root Harvesters:** Some machines are designed to harvest a variety of root crops, from potatoes to carrots to onions. These machines usually have adjustable components to suit different root crop types.
- **Self-Propelled Multi-Crop Harvesters:** Large-scale machines capable of harvesting multiple root crops. These are typically more expensive but save time and labor when growing diverse crops on the same land.

CLASSIFICATION OF ROOT HARVESTING EQUIPMENT

The root harvesting equipment can be classified in several different ways as described below:

1. According to the Source of Power

- Manual operated, viz. hand tools and digging aids such as spade, khurpa, digging and picking fork etc.
- Animal-drawn equipment such as potato-plough, animal drawn groundnut digger etc.
- Tractor operated equipment such as spinner-digger, elevator-digger, groundnut-digger-shaker-windrower, sugarbeet-harvester, carrot-harvester, radish-harvester etc.
- Self-propelled machine such as potato-harvester/combine, groundnut combine, sugarbeet harvester etc.

2. According to the Mode of Hitching

- Trailed type
- Mounted type

- Semi-mounted type
- Self-propelled type

3. According to the Unit Operations Performed:

- Simple digger
- Digger-shaker
- Digger-elevator-windrower
- Digger-shaker-conveyor/digger-shaker-windrower.

4. According to Completion of Operation:

- Direct harvesting equipment, i.e. equipment which accomplishes the complete operation of digging and separating to the soil as well as of collecting the recoverable products in one 'go'.
- Indirect harvesting equipment, i.e. equipment which accomplish the complete operation in more than one passes. For instance, a groundnut digger shaker windrower digs the vines and deposits them in a windrow after shaking out the soil. The vines are sun-dried for a few days and thereafter a groundnut combine with a pick-up attachment is used to pick the vines and separate the nuts.

Considerations for Proper Field Operation of Root Crop Harvesting Equipment

Following are some of the general considerations for efficient utilization of the root harvesting equipment:

- Proper field layout
- Proper planting method
- Proper moisture content at the harvest time
- Pre-harvest treatment such as haulm-cutting, vine-killing etc.
- Proper adjustment of the depth of the digging.
- Proper hitching of the machine.
- Proper adjustment of the speeds of the picking, conveying and shaking system
- Selection of desired forward speed in keeping with the field and crop conditions.
- Number of workers in the crew to pick and collect the material simultaneously.
- Matching of the transport and handling equipment to the capacity of the digger/harvester.
- Proper harvesting, maintenance and repairs of the equipment.

General Design Considerations for Root Crop Harvesting Equipment

For designing a root crop harvesting equipment, the following points should be kept in view:

- **Variety of the crop and its cultural practices**: In case the existing varieties and cultural practices pose some problems in mechanical harvesting, attempt should be made to make necessary modifications.

- **Soil Type**: The soil type and moisture content at the time of harvesting are important factors influencing the machine performance. Suitable soil-amending substances may help to develop the proper soil-texture to eliminate problem of excessive blade clogging and clod-formation. The soil for root crops should be friable for easy separation.
- **Method of Irrigation:** One must carefully consider the method of irrigation being followed. Depending upon where the crop is sown on ridges/beds/flat surface, proper method of irrigation should be selected. The availability of right moisture content at the time of harvesting is very essential for proper penetration of blade and efficient soil-fragmentation.

  Machine design considerations:
  - Type of the digging shovel, shape and size.
  - Conveying and separation system. It must be efficient to handle and separate the loose material with minimum injuries.
- **The height of fall**: Avoid excessive height of fall of the roots/tubers to lesson injuries. If required use inside rubber padding at the components to minimize external/internal injuries.
- **Proper speeds**: The speeds of the components used for the translocation and separation of the material should be such that these do not lead to excessive agitation, damage and loss of the products.
- **Controls:** The various controls for raising and lowering of the blade, varying speeds, etc should be carefully designed. Provision should be made for adjusting the speed of the components in accordance with the field conditions.
- Use of Multi-speed gear box and vary-speed drives is desirable on root crop harvesting machines.

POTATO-DIGGER

The mechanization of potato cultivation, particularly on larger farms, became necessary due to the labor-intensive nature of operations like sowing and harvesting. Manually harvesting potatoes requires approximately 800 man-hours per hectare, making it a highly labor-demanding task. The introduction of potato diggers has mechanized the digging process. These machines can be manually operated or powered, including animal-drawn, tractor-drawn, potato digger shakers, and potato digger elevators. The potato digger elevator consists of a shovel, rod chain conveyor, gearbox, and two gauge wheels. The machine digs one or two ridges of potatoes at a time and transfers the soil-potato mixture to the rod chain conveyor. The conveyor's agitating sprocket moves the rod chain, separating the soil. The tubers, with minimal soil, fall to the ground in a row behind the machine, where they are later collected manually. These machines are widely used in northern India and are commercially available. Most models are two-row machines, though both one-row and two-row options exist. The machine can cover 1.5-3.0 hectares per day at a forward speed of 2.0-3.0 km/h. Potatoes are susceptible to bruising or skinning, although fully mature potatoes are less vulnerable compared to immature ones.

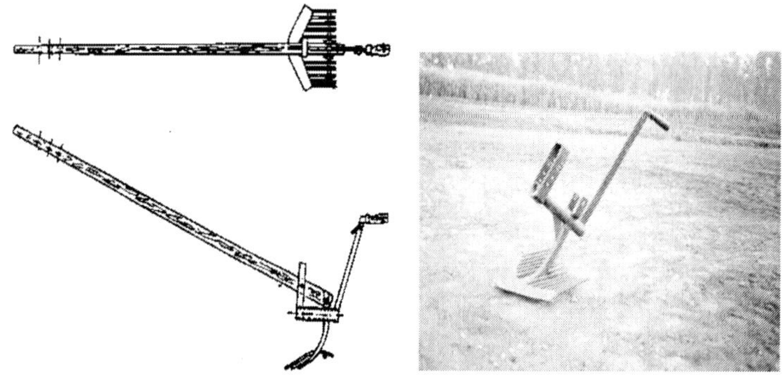

Figure 12.1: Animal-drawn Potato Digger

Figure 12.2: Tractor-operated 2-row Potato Digger

Figure 12.3: Tractor-operated Single Row Potato Digger Elevator

GROUNDNUT DIGGER SHAKER

The combination digger shaker consists of a single curved blade with rear extension rods, two disc coulters, and an endless chain-type pick-up elevator-cum-shaker conveyor (rattler) equipped with spiked rakes attached to the chains. Power is transmitted through a gearbox, cross drive shaft, V-belt & pulleys, and driving sprockets mounted on a shaft supported by the frame.

The blade, shanks, coulter forks, and gearbox are secured to a rectangular toolbar, which also has two inclined side members supporting the sprocket drive shaft. The shaker conveyor roller chains are mounted over two idler sprockets at the front end. Gauge wheels are included to regulate the working depth. The front portion of the shaker conveyor can be adjusted using a hand-operated lower chain. The cutting width is 1.22 meters.

Operating Principle of Groundnut Digger Shaker

The blade cuts a 1.2-meter-wide strip of crop. The vines are picked up and lifted by the shaker conveyor following the blade. The soil and debris on the roots and pods are shaken off, and the vines are deposited as a fluffy windrow at the rear. This placement exposes the pods to the sun. After 1-2 days of sun-drying, the vines are manually gathered and transported for threshing. In some countries, combine harvesters are used for similar operations.

The tractor PTO-operated groundnut digger shaker is widely used in northern India for groundnut harvesting. The machine includes a blade, elevator-cum-pick-up reel, power transmission system, windrower attachment, fenders, gauge wheels, and coulters. The pick-up reel's front end is adjustable to match the blade's working depth.

It digs up the plants along with the groundnuts. The material is moved over the conveyor, loosening and removing the soil. Designed for tractors with 30 hp or more, it is powered by the PTO shaft. The digging shovel may either be pointed or have a conveying cutting edge, with a width of 50-60 cm. This machine has an average output of 0.27 ha/h and operates at a forward speed of 2.4-3.0 km/h. To continuously operate the digger, about 10-12 workers are needed to pick up and gather the tubers from the windrow.

Groundnut digger blades with a corrugated roller at the rear are used in medium and heavy soils, as the roller helps crush clods, making tuber picking easier. In light soils where clod formation is minimal, a simpler digger blade is used, allowing loose soil to fall through the extension rods, leaving groundnut tubers on the soil for easy collection.

Final Thoughts

Root harvesting equipment is essential for modern farming, allowing farmers to harvest root crops efficiently and with minimal damage to the produce. From small, tractor-pulled harvesters to large, self-propelled machines, root harvesters offer solutions for different scales of farming, ensuring high productivity and better crop management. Proper maintenance and operation of these machines are essential to maximizing their benefits.

13

Construction and Function of Threshing Equipment

THRESHING EQUIPMENT

A thresher is a machine designed to separate grains from harvested crops while minimizing loss and damage. During threshing, various types of grain loss may occur, such as broken, un-threshed, blown, or spilled grains.

"Threshing" refers to the process of separating grains from plants. This operation can take place in the field or on the threshing floor, using hand labor, animals, or machines. Regardless of the method used, it's crucial that threshing is done carefully to avoid grain or husk breakage, which can lower the quality of the product and increase losses from insects or mold. Extra care should be taken when transporting harvested crops to the threshing area to prevent any loss.

Threshing or Shelling Operations

Threshing or shelling operations occur after harvesting and any pre-drying processes. These tasks can be performed either in the field or at the farm, using hand labor, animals, or machines.

The method of threshing or shelling is influenced by various agronomic, economic, and social factors. It can be done in several ways:

- By hand, using simple tools.
- With the help of animals or vehicles.
- Using manually operated mechanical machines.
- Using motorized equipment for mechanical threshing or shelling.

As noted earlier, harvesting and threshing or shelling can often be done together using combine-harvesters or picker-shellers.

Regardless of the system used, it's essential to handle threshing or shelling with care. Improper methods can cause grain or husk breakage, reducing product quality and leading to higher losses from insects and mold. Additionally, transporting the product from the field to the threshing or shelling location requires careful attention to prevent significant losses.

Hand Threshing

Hand threshing is commonly used for pigeonpea. One of the simplest methods is striking the sheaves of the crop, spread over a threshing floor, with a flail or stick. While this technique can also be used for other pulse crops such as chickpea, mungbean, urdbean, fieldpea, rajmash, and lentil, threshing with animals or vehicles is more convenient. The threshing floors must have a hard, clean surface for this method. A worker can thresh between 15-40 kg of product per hour using hand-threshing.

Another common method for threshing rice is trampling it underfoot. Similarly, threshing of rice, sorghum, beans, and groundnuts can be done by striking sheaves on a threshing floor with a flail or stick. The threshing floors need to have a hard, clean surface. This hand-threshing method also yields 15 to 40 kg per hour, depending on the worker.

Threshing with Animals or Tractors

When draught animals are available, and there is a large quantity of rice, threshing can be done by driving the animals (harnessed to threshing devices) over sheaves about 30 cm thick. This operation, known as "treading out," can also be done with vehicles. In some Asian countries, tractors are used for threshing rice instead of draught animals.

To thresh rice with a tractor, the sheaves are spread in layers on a circular threshing-floor, typically 15-18 meters in diameter. The tractor passes twice over the sheaves, with the sheaves being turned over between passes. Alternating between two contiguous threshing-floors can yield up to 640 kg per hour.

Maize-Shelling with Hand-Operated Machines

Manual shellers, often made by local artisans, facilitate faster and easier shelling of maize ears. These shellers come in various models, some of which can be motorized, and are generally operated by hand or pedal. Typically, one worker is sufficient to use the manual sheller. Yields range from 14 to 100 kg per hour, making these machines suitable for small-scale production needs.

Threshing or Shelling with Motorized Equipment

Threshing or shelling with motorized equipment primarily refers to the use of motorized threshing machines. While they are increasingly being replaced by combine harvesters, these machines still play a significant role in the post-harvest process due to their versatility. By simply changing a few accessories and adjusting settings, these machines can be used for various types of grain, including rice, maize, sorghum, beans, sunflowers, wheat, and soybeans. They typically feature a rotating threshing drum with beaters or teeth and a stationary counter-thresher, along with mechanisms to shake out the straw, clean, and bag the grain.

Motorized threshers, whether self-propelled or tractor-drawn, are often mounted on rubber tires for easy transport to the field. Operating these machines generally requires two or three workers, and the yields can vary from 100 to 5,000 kg per hour, depending on the machine type, grain characteristics, worker skill, and organization of the work.

MULTI -CROP THRESHER

Features

The thresher includes a threshing cylinder, oscillating box, straw walker, and a winnowing and cleaning attachment. The rasp-bar cylinder is constructed from cast iron rings, sheet metal, and toothed racks, mounted on the main shaft supported by two heavy pedal bearings. Several pulleys of varying sizes are attached to this shaft to transmit power to the winnowing and fan system. The machine allows for adjustments to the cylinder and blower speeds, as well as the concave clearance, making it adaptable for different types of crops.

Uses

Suitable for threshing paddy, ragi, jowar, maize, sunflower, wheat etc.

AXIAL FLOW PADDY THRESHERS

Features

The thresher is equipped with a spike-tooth threshing cylinder, concave, cylinder casing, cleaning system, and feeding chute. In the axial flow design, the crop is fed at one end, and after thorough threshing, the straw is expelled from the opposite end. During the threshing process, the crop rotates three and a half times around the cylinder, ensuring all grains are detached. The threshing cylinder is of the peg type, and the casing features seven louvers to facilitate axial movement of the crop. The cleaning system includes two aspirator blowers and two sieves.

GROUNDNUT THRESHER

Features

The thresher operates on an axial flow design and includes a feed hopper, spike-tooth cylinder, concave, oscillating sieves, and a blower. The cylinder features 10 rows of pegs and is enclosed by a wire mesh concave with sieve openings measuring 80 x 25 mm. Below the cylinder and concave assembly, two oscillating sieves are installed to separate pods from leaves, soil, and dust particles. A blower positioned between the sieves helps expel the leaves. This thresher is also suitable for freshly harvested crops with green vines.

Uses:

Separation of groundnut pods from harvested plants.

MAIZE DE-HUSKER/SHELLER

Features

Maize dehusker cum sheller are of two types namely, spike tooth type and axial type. In the spike tooth type sheller, pegs are staggered at varying heights for better shelling efficiency. The spikes are placed in 6 rows with 6 pikes in each row. The sieves have 1.25 cm diameter opening to separate the shelled maize from husk. In axial flow type threshers, pegs are provided on the cylinder and louvers were provided on the upper periphery of the drum to convey the crop to the outlet.

Uses

It is used for dehusking and shelling of maize cobs simultaneously.

COMPONENTS OF THRESHER

The thresher machine comprises several components, including a fan/blower, spikes, flywheel, concave, cylinder, feeding chute, towing hook, frame, driver pulley, upper sieve, transport wheel, can pulley, shutter plate, and suspension lever.

Key Components of a Thresher

- **Concave:** A concave-shaped metal grating partially surrounds the threshing cylinder. It helps rub grains from the plant or ear heads against the cylinder, allowing the grains to fall through onto the sieve.
- **Threshing Cylinder:** A rotating mechanism that separates the grains from the stalks by beating and rubbing them against the concave.
- **Cleaning Unit:** Ensures the removal of chaff and other impurities, leaving only clean grains for collection.

Threshing Cylinder

- Most important component of thresher
- Balanced rotating assembly comprising rasp beater bar or spikes on its periphery and their support for threshing the crop
- Types
    - Peg tooth
    - Wire loop
    - Rasp bar
    - Angle bar
    - Hammer mill

Types of Threshing Cylinder

Peg tooth

- The teeth on the concave & cylinder are so arranged that the cylinder teeth pass midway between the staggered teeth on the concave
- The clearance between the cylinder & the concave is adjusted according to the requirement
- As the stalks pass thru the clearance space, the grains get separated from the head due to impact action between the teeth

Wire loop

- Cylinder is studded with number of wire loops through out its outer periphery
- Mostly used on paddy thresher

Angle bar

- Cylinder is equipped with angle iron bars, helically fitted on the cylinder
- The bars have rubber pads on their faces
- The clearance between cylinder and concave unit at the entrance is from 13 mm to 19 mm and reduces to 6 to 9 mm only

Hammer mill type

- Beaters are in the shape of hammer mill
- Beaters are attached with the beater arm at the tip
- Beater arms are rigidly fixed to a hub which is mounted on main shaft

Rasp bar cylinder

- Cylinder has corrugated bars round it
- Threshing is accomplished between corrugated cylinder bars and stationary bars of the concave portion
- Rotating cylinder takes the grains out from the head as it is drawn over the bars on the concave unit
- Usually 6 to 8 bars are spirally fixed on the cylinder

Cleaning unit

- Function is to separate & clean the threshed grain
- Mainly consists of two or more oscillating sieves, a fan and air sucking duct known as aspirator
- Usually two ducts viz. primary and secondary duct
- Function of primary duct is to remove major portion of straw, dust and other foreign matter
- Secondary duct is used for final cleaning of the grain

Combine Harvester Thresher

- Machine designed for harvesting, threshing, separating, cleaning and collecting grains while moving through the standing crop
- Main functions are
    - Cutting the standing crops
    - Feeding the crop to threshing unit
    - Threshing the crops
    - Cleaning the grains from straw
    - Collecting the grains in a container

PROCESS OF THRESHER

Threshing machines are widely used in harvesting and threshing operations, involving a three-step process:

First Step

The process begins by feeding straw and grain bundles into the feeder, which regulates the feed rate to prevent overloading the machine.

Second Step

The separator, a rapidly rotating set of blades, takes over. It separates the straw and grains by tearing the bundles apart, breaking twine, and detaching the heads from the straw. The straw is then beaten against a grooved plate and passed over a straw rake to remove most of the straw from the kernels.

Third Step

Finally, the clean kernels pass through a screen and are subjected to an air stream that blows away any remaining chaff and straw. The pure grains fall into a feeder with a measuring device, while straw is expelled onto a stack by a powerful blower.

Safety Precautions While Operating a Threshing Machine

- Avoid feeding ear-heads without stalks to prevent hand injuries.
- Ensure the area is free of obstructions during operation.
- Refrain from wearing loose clothing, bangles, or wristwatches near the thresher.
- Set adequate lighting for nighttime operations to avoid accidents.

Bibliography

Anbari, S., A. Lundkvist and T. Verwijst. 2011. Sprouting and shoot development of Sonchus arvensis in relation to initial root size. Weed Research 51: 142–150.

Ascard, J. 1994. Dose-response models for flame weeding in relation to plant size and density. Weed Research 34: 377–385.

Ascard, J. 1995a. Thermal weed control by flaming: biological and technical aspects. Institutionen für lantbruksteknik, (Swedish University of Agricultural Sciences), Alnarp Report 200: 1–61.

Ascard, J. 1995b. Effects of flame weeding on weed species at different developmental stages. Weed Research 35: 397–411.

Buhler, D.D., K.A. Kohler and R.L. Thompson. 2001. Weed seed bank dynamics during a five-year crop rotation. Weed Technology 15: 170–176.

Clements, D.R., S.E. Weise, R. Brown, D.P. Stonehouse, D.J. Hume and C.J. Swanton. 1995. Energy analysis of tillage and herbicide inputs in alternative weed management systems. Agriculture, Ecosystems and Environment 52: 119–128.

Cussans, G.W., S. Raudonius, P. Brain, and S. Cumberworth. 1996. Effects of depth of seed burial and soil aggregate size on seedling emergence of Alopecurus myosuroides, Galium aparine, Stellaria media and wheat. Weed Research 36: 133–141.

Gallagher, R.S. and J. Cardina. 1998. Phytochrome mediated Amaranthus germination II: development of very low fluence sensitivity. Weed Science 46: 53–58.

Green, M.B. 1987. Energy in pesticide manufacture, distribution and use. In: Energy in Plant Nutrition and Pest Control, eds. B.A. Staut and M.S. Mudahar. pp. 165–177. Elsevier: Amsterdam.

Grubinger, V. and M.J. Else. No date. Vegetable Farmers and their Weed-Control Machines. DVD 75 min.

Mirsky, S.B., E.R. Gallandt, D.A. Mortensen, W.S. Curran and D.L. Shumway. 2010. Reducing the germinable weed seed bank with soil disturbance and cover crops. Weed Research 50: 341–352.

Mohler, C.L., J.C. Frisch and C.E. McCulloch. 2006. Vertical movement of weed seed surrogates by tillage implements and natural processes. Soil and Tillage Research 86: 110–122.

Mohler, C.L., J.C. Frisch and J. Mt. Pleasant. 1997. Evaluation of mechanical weed management programs for corn (Zea mays). Weed Technology 11: 123–131.

Mohler, C.L., M. Liebman and K. Staver. 2001. Mechanical management of weeds. In Liebman, M., C.L. Mohler and C.P. Staver. 2001. Ecological Management of Agricultural Weeds. Cambridge University Press: New York.

Mohler, C.L. 1993. A model of the effects of tillage on emergence of weed seedlings. Ecological Applications 3: 53–73.

Monsanto. 2010. Roundup PowerMAX® Label. Monsanto Company: St. Louis, Missouri.

Mt. Pleasant, J. and R. Burt. 1994. Time of cultivation in corn: effects on weed levels and grain yields. Proceedings of the Northeastern Weed Science Society 48:66.

NCCE. 2008. Conversion Factors for Bioenergy. North Carolina Cooperative Extension.

OSU. 2005. Weed 'Em and Reap. Part I. Tools for Non-Chemical Weed Management in Vegetable Cropping Systems. Oregon State University, Department of Horticulture: Corvallis, OR. DVD 36 min.

Rahkonen, J., and P. Vanhala. 1993. Response of a mixed stand to flame and use of temperature measurements predicting weed control efficiency. International Conference I.F.O.A.M. Non-Chemical Weed Control Proceedings, Dijon: 177–181.

Sahay, J. (2008). Elements of Agricultural Engineering. Standard Publishers Distributors, New Delhi.

Singh, G. (2006). Farm Machinery: Principles and Applications. Jain Brothers, New Delhi.

Singh, G., and Varshney, A. C. (2010). Farm Power and Machinery Engineering. Jain Brothers, New Delhi.

TNAU. (2017). Practical Manual on Operation and Maintenance of Agricultural Machinery. Department of Agricultural Engineering, Tamil Nadu Agricultural University.

Index

I

J

L

M

O

P

R

S

T

V

W